T0305514

# Time-Frequency Analysis Techniques and Their Applications

Most of the real-life signals are non-stationary in nature. The examples of such signals include biomedical signals, communication signals, speech, earthquake signals, vibration signals, etc. Time-frequency analysis plays an important role for extracting the meaningful information from these signals. The book presents time-frequency analysis methods together with their various applications.

The basic concepts of signals and different ways of representing signals have been provided. The various time-frequency analysis techniques namely, short-time Fourier transform, wavelet transform, quadratic time-frequency transforms, advanced wavelet transforms, and adaptive time-frequency transforms have been explained. The fundamentals related to these methods are included. The various examples have been included in the book to explain the presented concepts effectively. The recently developed time-frequency analysis techniques such as, Fourier-Bessel series expansion-based methods, synchrosqueezed wavelet transform, tunable-Q wavelet transform, iterative eigenvalue decomposition of Hankel matrix, variational mode decomposition, Fourier decomposition method, etc. have been explained in the book. The numerous applications of time-frequency analysis techniques in various research areas have been demonstrated.

This book covers basic concepts of signals, time-frequency analysis, and various conventional and advanced time-frequency analysis methods along with their applications. The set of problems included in the book will be helpful to gain an expertise in time-frequency analysis. The material presented in this book will be useful for students, academicians, and researchers to understand the fundamentals and applications related to time-frequency analysis.

# Time-Frequency Analysis Techniques and Their Applications

Ram Bilas Pachori

## CRC Press
Taylor & Francis Group
Boca Raton London New York

CRC Press is an imprint of the
Taylor & Francis Group, an **informa** business

First edition published 2023
by CRC Press
6000 Broken Sound Parkway NW, Suite 300, Boca Raton, FL 33487-2742

and by CRC Press
4 Park Square, Milton Park, Abingdon, Oxon, OX14 4RN

*CRC Press is an imprint of Taylor & Francis Group, LLC*

ISBN: 978-1-032-39297-4 (hbk)
ISBN: 978-1-032-43576-3 (pbk)
ISBN: 978-1-003-36798-7 (ebk)

DOI: 10.1201/9781003367987

Typeset in Time new Roman
by KnowledgeWorks Global Ltd.

# *Dedication*

---

*Dedicated to my parents*

# Contents

# Foreword

The subject of time-frequency analysis is evergreen, both, from a theoretical perspective, as also, from the view of practical applications. The reasons are many. However, two of them stand out, to my mind: The first is, that the very origin of the subject is, in a very fundamental limitation that 'Nature itself imposes': in the form of the uncertainty principle, which limits simultaneous time-frequency localization to an insurmountable lower bound on the time- frequency spread of a signal. The second is, that there is no 'perfect' time-frequency analysis tool from an operational viewpoint and every approach to time-frequency analysis has its 'pros and cons'. Since time-frequency analysis lies at the heart of many practical applications in signal and image processing, this book, authored by Ram Bilas Pachori, which explores the subject in depth and presents its tenets in a manner that appeals to both, the beginner and the advanced student and researcher is, indeed, a valuable contribution to the academic community.

The organization of material in this scholarly work aptly fulfils the needs of many academicians and researchers. It 'starts at the very beginning', so to speak, describing the essential types and properties of signals and their analytical tools. It then leads the reader to more and more beautiful facets of the subject of time-frequency analysis, including the traditional short time Fourier transform, the wavelet transform, its variants and several advanced possibilities. What makes this book particularly valuable is its in-depth treatment of the Fourier-Bessel approach to time-frequency analysis. I believe this is a rare source on time-frequency analysis, in that respect: few expositions to the subject deal with this aspect in depth, whereas it actually deserves a 'place of pride' in this domain of knowledge. Ram Bilas Pachori has done well, to give this dimension of the subject, its much needed status of importance. I believe it stems from the author's own, unique command over and insight into, this aspect of time-frequency analysis, right from the days of his doctoral Thesis investigations.

Another important aspect of this book is its collection of rich problem sets, located at the end of each chapter. This makes the book valuable both to the teacher of the subject and, also to the student. Concepts have been explained with appropriate illustration and examples, which makes it easier to grasp the material. A later section, detailing a very comprehensive set of applications of the subject, adds special value to this book.

It is my proud privilege and pleasure, to introduce this wonderful contribution by Ram Bilas Pachori to the signal and image processing community. In fact, it will serve as a reference and text, to many a seeker of knowledge from other domains of engineering and science, as well, as the chapter on applications brings out very clearly. May this scholarly tome inspire many a mind and satisfy many a seeker of knowledge!

Vikram M. Gadre
Professor, Electrical Engineering, IIT Bombay

# Preface

From the domain of biomedical engineering to fluid dynamics, most of the signals being generated in the process possess a non-stationary nature. The events in these signals change with time hence, analysis of these signals using conventional Fourier analysis technique fails to give meaningful information to localize various events. In order to do so, time-varying spectral content is needed which can be obtained using time-frequency analysis (TFA) of the signal. This book connects the analysis of signals in time- and frequency-domains with joint time-frequency domain. This specific structure can be helpful to many undergraduate and postgraduate students, researchers, and academicians to enhance their knowledge for non-stationary signal processing. It also provides the brief insight on some recent applications of joint TFA techniques to various engineering problems. The content of the book is as follows:

1. Basic concepts of signals.

2. Concepts of TFA from basic to advanced.

3. Usefulness of TFA to various real-time engineering problems.

The organization of this book is as follows:

Chapter 1 introduces the basics of signals, which includes definition and classification of signals. Various important signals are also discussed which can be found to be helpful in analysis of real-time signals. At the end of this chapter, various types of signal operations are discussed which can be performed on a signal during analysis.

Chapter 2 presents the idea of representation of signals in terms of some fundamental or basic set of orthogonal signals. Least square interpretation of the signal representation is also discussed. Concept of finality of coefficients is discussed for signal representation in terms of orthogonal basis functions. Concept of sampling is discussed as a signal representation in terms of impulses. Also, Fourier analysis techniques are presented for frequency-domain analysis of periodic and aperiodic signals. Further, Fourier-Bessel series expansion (FBSE) is explained in detail for the analysis of non-stationary signals.

Chapter 3 explains the steps involved in various time-domain and frequency-domain signal characteristics like, mean time, duration, mean frequency, and frequency spread. Using these signal characteristics, the uncertainty principle and Heisenberg box representation are discussed for the signals localized in both time- and frequency-domain. Also, for amplitude-frequency modulated (AFM) signals, the extraction of amplitude envelope (AE) and instantaneous frequency (IF) functions using discrete energy separation algorithm (DESA) and Hilbert transform separation algorithm (HTSA) is explained. The HTSA-based Hilbert spectral analysis (HSA) is also explained for obtaining the time-frequency representation (TFR) of a signal.

Chapter 4 discusses the short-time Fourier transform (STFT) for the analysis of non-stationary signals. Various interpretations of STFT are discussed along with its various properties. Filter bank structure of STFT is discussed. Short-frequency Fourier transform (SFFT) and its squared magnitude, known as sonogram are discussed.

Chapter 5 presents the theory of wavelet-based analysis of non-stationary signals along with its various properties. Energy distribution in time-scale plane, i.e., scalogram of a signal using wavelet transform is presented. Also, the concept of multiresolution for discrete wavelet transform (DWT) is discussed. A filter-bank-based DWT is presented at the end of this chapter.

Chapter 6 focusses on various quadratic time-frequency distributions (TFDs) for a non-stationary signal. Techniques covered in this chapter are Wigner-Ville distribution (WVD), pseudo WVD, smoothed pseudo WVD, ambiguity function, Cohen's class TFDs, WVD, Choi-Williams distribution, and spectrogram as special cases of Cohen's class TFDs, and affine-invariant TFDs.

Chapter 7 explains the various advanced wavelet transforms for analysis of non-stationary signals like, wavelet packet transform (WPT), synchrosqueezed wavelet transform (SWT), rational-dilation wavelet transforms (RDWT), tunable-Q wavelet transform (TQWT), flexible analytic wavelet transform (FAWT), FBSE-based FAWT (FBSE-FAWT), and dual-tree complex wavelet transform (DTCWT).

Chapter 8 introduces the data-adaptive signal decomposition techniques. The techniques covered in this chapter for decomposition of signals into AFM signals are empirical mode decomposition (EMD), ensemble EMD (EEMD), variational mode decomposition (VMD), empirical wavelet transform (EWT), FBSE-based EWT (FBSE-EWT), Fourier decomposition method (FDM), iterative eigenvalue decomposition of Hankel matrix (IEVDHM), and dynamic mode decomposition (DMD). The use of HSA is also presented to obtain TFR of the signals using the aforementioned decomposition techniques.

Chapter 9 presents various applications of techniques considered in previous chapters. Various signals considered in this chapter are biomedical signals and images, speech signals, communication signals, power-quality signals, financial signals, vibration signals, and signals occurred in chemical engineering and ocean engineering.

At the end of every chapter, a set of conceptual problems is provided which might be helpful to build a strong foundation in the subject.

# Acknowledgments

The inspiration to write this book comes from the time of my teaching at the International Institute of Information Technology Hyderabad, India where I offered a course on Time-Frequency Analysis for senior undergraduate and postgraduate students. Many of the students in my class expressed a need for a book on Time-Frequency Analysis which could be linked with Signals and Systems. I recognized the need for a book in this area and started to prepare lecture notes and material from a book writing point of view. I continued teaching Time-Frequency Analysis course at Indian Institute of Technology Indore, India and updated the course material. This prepared course material became the first draft of the book during the COVID-19 lockdown.

I am very thankful to many mentors and teachers who have provided inspiration and knowledge during my career. I am very grateful to Professor Pradip Sircar for introducing me to the domain of non-stationary signal processing during my M.Tech. and Ph.D. programs at Indian Institute of Technology Kanpur, India. Professor David Hewson, Professor Hichem Snoussi, and Professor Jacques Duchêne provided a lot of exposure to important applications of non-stationary signal processing in the area of biomedical signal processing during my Post-doctoral research at University of Technology of Troyes, France. I also learned about brain-computer interfacing from Professor Girijesh Prasad at Ulster University, Londonderry, UK during my time as a visiting scholar at the same university.

I am very grateful to my students who provided feedback, suggestions, and assessment during teaching the Time-Frequency Analysis course, which helped a lot to improve the quality of the course material and also the content of the book. I would also like to gratefully acknowledge the help and support provided by my colleagues at Indian Institute of Technology Indore, India. I would like to convey my thanks to Dr. Nitya Tiwari for her valuable suggestions.

I am also grateful to my present and past students at Signal Analysis Research Lab (SARL), Indian Institute of Technology Indore, India for their very useful suggestions, comments, and help in completing this book. I would like to convey my special thanks to my students, Kritiprasanna Das, Vivek Kumar Singh, Pradeep Kumar Chaudhary, Shailesh Bhalerao, Aditya Nalwaya, Ashok Mahato, Achinta Mondal, Nabasmita Phukan, Shailesh Mohine, Vaibhav Mishra, Bethapudi Shirly Susan, Akah Precious C, Amishi Vijay, and Alavala Siva Sankar Reddy for their constructive suggestions to improve the book.

I also express my thanks to the Senior Editor, Mr. Gauravjeet Singh Reen and reviewers for their useful comments and suggestions to improve this book.

Finally, I would like to thank my family members for their kind support during the writing of this book. I am especially grateful to my father and son who reminded me many times to complete this book. Last but not least, I would like to thank and praise Lord Krishna for granting countless blessings and the opportunity to write this book.

Ram Bilas Pachori

# Author Biography

**Ram Bilas Pachori** received the B.E. degree with honors in Electronics and Communication Engineering from Rajiv Gandhi Technological University, Bhopal, India in 2001, the M.Tech. and Ph.D. degrees in Electrical Engineering from Indian Institute of Technology Kanpur, India in 2003 and 2008, respectively.

He worked as a Post-Doctoral Fellow at Charles Delaunay Institute, University of Technology of Troyes, France during 2007–2008. He served as an Assistant Professor at Communication Research Center, International Institute of Information Technology, Hyderabad, India during 2008–2009. He served as an Assistant Professor at Department of Electrical Engineering, Indian Institute of Technology Indore, India during 2009–2013. He worked as an Associate Professor at the Department of Electrical Engineering, Indian Institute of Technology Indore, during 2013–2017, where presently he has been working as a Professor since 2017. Currently, he is also associated with Center for Advanced Electronics at Indian Institute of Technology Indore. He was a Visiting Professor at Neural Dynamics of Visual Cognition Lab, Free University of Berlin, Germany during July-September, 2022. He has served as a Visiting Professor at School of Medicine, Faculty of Health and Medical Sciences, Taylor's University, Malaysia during 2018–2019. Previously, he has worked as a Visiting Scholar at Intelligent Systems Research Center, Ulster University, Londonderry, UK during December 2014.

His research interests are in the areas of Signal and Image Processing, Biomedical Signal Processing, Non-stationary Signal Processing, Speech Signal Processing, Brain-Computer Interfacing, Machine Learning, and Artificial Intelligence and Internet of Things in Healthcare.

He is an Associate Editor of Electronics Letters, IEEE Transactions on Neural Systems and Rehabilitation Engineering, Biomedical Signal Processing and Control and an Editor of IETE Technical Review. He is a senior member of IEEE and a Fellow of IETE, IEI, and IET.

He has 284 publications which include journal papers (173), conference papers (78), books (10), and book chapters (23). He has also eight patents: 01 Australian patent (granted) and 07 Indian patents (filed). His publications have been cited approximately 13000 times with h-index of 62 according to Google Scholar (March 2023). He has worked on various research projects with funding support from SERB, DST, DBT, CSIR, and ICMR.

# 1 Basics of Signals

*"Arise awake and stop not till the goal is reached."* –Swami Vivekananda

## 1.1 DEFINITION OF SIGNAL

Signal can be defined as a function of independent variables which is able to convey information. In order to have meaningful interpretation, the signals are considered as an output of the system and similar way, the system is defined with the help of input and output signals. In this way, signals and systems are dependent on each other. Based on the number of independent variables used in the functional representation of the signal, the signal can be one-dimensional (1D) like a speech signal, two-dimensional (2D) like an image, and three dimensional (3D) like a video, etc. Signals contain information about the system from which they have been generated [1]. For example, electrocardiogram (ECG) signals measure electrical activity of the heart. Similarly, electroencephalogram (EEG) signals measure electrical activity of the brain. There are many examples of signals, which include speech signals, biomedical signals such as EEG, ECG, magnetic resonance imaging (MRI), computerized tomography (CT) scan, fundus images, communication signals, images and video signals, seismic signals, vibration signals, etc. The signals can be represented with the help of functions which contain independent variables. The function is represented as a relation between two sets (independent and dependent) which follows either one-to-one or many-to-one property. Figure 1.1 (a) shows the graphical representation of function from the relation. Such functions with information provide signal representation [2].

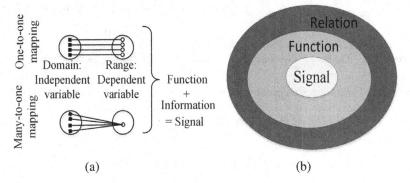

|  |  |
|---|---|
| (a) | (b) |

**Figure 1.1** (a) Functional representation of the signal and (b) syllogism representation of relation, function, and signal.

It should be noted that all relations are not necessarily functions. The relations that satisfy any of the one-to-one properties and many-to-one properties are known

DOI: 10.1201/9781003367987-1

as functions. Similarly, all functions are not necessarily signals, only those functions that are able to convey information are known as signals. The relation between signal, function, and relation is shown by Venn diagram in Fig. 1.1 (b). The relation, function, and signal can be defined as follows:

> Relation: It is characterized by one-to-one, one-to-many, many-to-one, or many-to-many mappings between domain (independent) and range (dependent) sets.

> Function: It has a one-to-one mapping or many-to-one mapping between two sets, domain (independent) and range (dependent).

> Signal: This is a function with ability to convey information.

Based on these domain- and range-based representations, the 1D and 2D signals are shown in Fig. 1.2. The mapping between domain ($t$) and range ($x$) can be

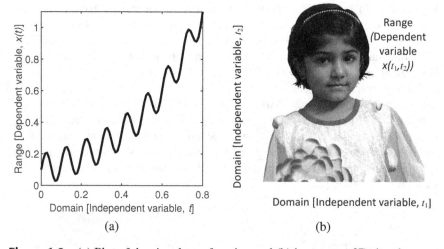

**Figure 1.2**    (a) Plot of the signal as a function and (b) image as a 2D signal.

represented as a Cartesian product of set $x$ and set $t$ in the form of relation as follows:

$$\{t\} \times \{x\} = \{t_1, t_2, t_3, ..., t_N\} \times \{x_1, x_2, x_3, ..., x_N\} \tag{1.1}$$

$$\{t\} \times \{x\} = \left\{ \begin{array}{l} (t_1, x_1), (t_1, x_2), (t_1, x_3), ..., (t_1, x_N) \\ (t_2, x_1), (t_2, x_2), (t_2, x_3), ..., (t_2, x_N) \\ \qquad\qquad\quad \cdot \\ \qquad\qquad\quad \cdot \\ \qquad\qquad\quad \cdot \\ (t_N, x_1), (t_N, x_2), (t_N, x_3), ..., (t_N, x_N) \end{array} \right\} \tag{1.2}$$

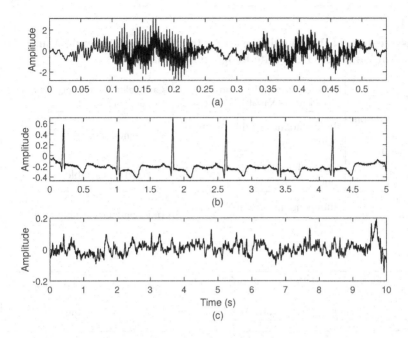

**Figure 1.3** Different real-time signals: (a) speech signal, (b) ECG signal, and (c) EEG signal.

This relation form can be expressed in term of function form as given by,

$$\{t\} \times \{x\} = \{(t_1, x_1), (t_2, x_2), (t_3, x_3), ..., (t_N, x_N)\} \tag{1.3}$$

Here, $x_1, x_2, x_3, \ldots, x_N$ can be of different values or have same value. Generally, signal can be interchanged by function because all signals are functions but the reverse is not true. In many cases, this independent variable (domain) is time and dependent variable (range) is amplitude and signal can be represented by $x(t)$, where $t$ is the domain and $x(t)$ is the range.

Such examples include speech, ECG, and EEG signals. The plots of speech, ECG, and EEG signals are shown in Fig. 1.3 (a)–(c), which have been obtained from the MATLAB toolbox mtlb.mat, Physionet database (MIT-BIH arrhythmia database) [3, 4], and the Bonn University EEG database [5], respectively.

## 1.2  TYPES OF SIGNALS

### 1.2.1  CONTINUOUS-TIME, DISCRETE-TIME, AND DIGITAL SIGNALS

Based on the nature of dependent and independent variables of the signals, they can be categorized into three categories namely, continuous-time signals, discrete-time signals, and digital signals [6]. Such signals are explained in Table 1.1.

**Table 1.1**

**Classification of the Signals Based on the Nature of Dependent and Independent Variables Over Finite Range and Domain**

| Category | Nature of Dependent and Independent Variables | Cardinality of Independent and Dependent Variables |
|---|---|---|
| Continuous-time signals | Independent variable : Continuous | Independent variable : ∞ |
| | Dependent variable : Continuous | Dependent variable : ∞ |
| Discrete-time signals | Independent variable : Discrete | Independent variable : Finite |
| | Dependent variable : Continuous | Dependent variable : ∞ |
| Digital signals | Independent variable : Discrete | Independent variable : Finite |
| | Dependent variable : Discrete | Dependent variable : Finite |

The examples of continuous-time, discrete-time, and digital signals are shown in Fig. 1.4 (a)–(c), respectively.

### 1.2.2   APERIODIC AND PERIODIC SIGNALS

The periodic signals satisfy the following condition:

$$x(t) = x(t+T), \quad \forall t \tag{1.4}$$

Where $T$ is a fundamental period of the signal $x(t)$. On the other hand, the aperiodic signals do not satisfy the above-mentioned mathematical relation.

The examples for periodic and aperiodic signals have been depicted in Figs. 1.5 (a) and (b), respectively.

### 1.2.3   ODD AND EVEN SIGNALS

An odd signal satisfies the following mathematical relation:

$$x(t) = -x(-t) \tag{1.5}$$

On the other hand, the even signal satisfies the following condition:

$$x(t) = x(-t) \tag{1.6}$$

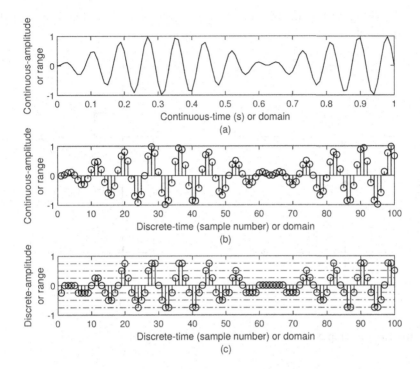

**Figure 1.4**  Example of (a) continuous-time signal, (b) discrete-time signal, and (c) digital signal.

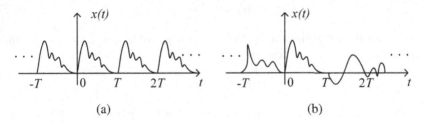

**Figure 1.5**  Example of (a) periodic signal and (b) aperiodic signal.

It should be noted that any signal $x(t)$ can be represented as a sum of even and odd signals, i.e.,

$$x(t) = \underbrace{\frac{x(t)+x(-t)}{2}}_{\text{Even signal part}} + \underbrace{\frac{x(t)-x(-t)}{2}}_{\text{Odd signal part}} \quad (1.7)$$

The above representation points out that for even signals, the odd part of the signal is zero. Similarly, in case of odd signals, the even part of the signal will be zero. An example of even signal ($x_e(t)$) and odd signal ($x_o(t)$) is shown in Figs. 1.6 (a) and (b), respectively.

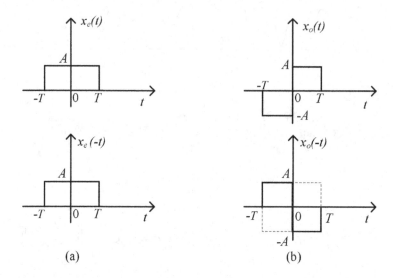

**Figure 1.6**　Example of (a) even signal and (b) odd signal.

### 1.2.4　CAUSAL, NON-CAUSAL, AND ANTI-CASUAL SIGNALS

The causal signals are defined as,

$$x(t) = 0, \quad t < 0 \tag{1.8}$$

It means these signals have zero values for negative time instants. On the other hand, the anti-causal signals can have non-zero values at negative time instants, i.e.,

$$x(t) = 0, \quad t > 0 \tag{1.9}$$

Non-casual signal may have non-zero values in both positive and negative time instants. Figures 1.7 (a), (b), and (c) show an example of causal, non-causal, and anti-casual signals, respectively.

### 1.2.5　DETERMINISTIC AND STOCHASTIC SIGNALS

Deterministic signals can be represented by mathematical models, rules, algorithms, etc., whereas stochastic signals can be studied with the help of probabilistic approaches. Examples of deterministic and stochastic signals can be seen in Figs. 1.8 (a) and (b), respectively. Figure 1.8 (a) is a sinusoidal signal, $x(t) = \sin(16\pi t)$. Stochastic signal, shown in Fig. 1.8 (b), is generated using the MATLAB's inbuilt function 'rand' to generate a signal whose values are from uniform distribution.

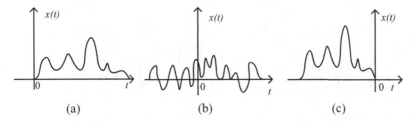

**Figure 1.7** Example of (a) casual, (b) non-casual, and (c) anti-casual signals.

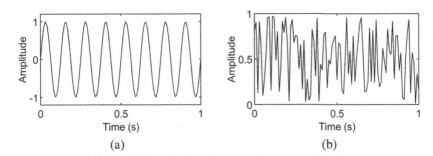

**Figure 1.8** Example of (a) deterministic signal and (b) stochastic signal.

### 1.2.6 STATIONARY AND NON-STATIONARY SIGNALS

In signal processing, the mathematical models are used to represent the signals. If the parameters of these models are constant with respect to the time, then such kind of signals are known as stationary signals. If any of the parameters of these models changes with respect to time then such signals are known as non-stationary signals. For example, sinusoidal signal and amplitude modulated sinusoidal signal are shown in Figs. 1.9 (a) and (b) for representing stationary and non-stationary signals, respectively. The stationary signal which is shown in Fig. 1.9 (a) has constant amplitude and constant frequency parameters. For stationary signal, a simple sinusoidal signal, $x(t) = \sin(30\pi t)$, with frequency of 15 Hz has been considered. On the other hand, the non-stationary signal is shown in Fig. 1.9 (b), has time-varying amplitude which can be written mathematically as, $x(t) = [1 + 0.5\sin(6\pi t)]\sin(30\pi t)$.

### 1.2.7 SINGLE-CHANNEL AND MULTI-CHANNEL SIGNALS

The signal associated with one variable is termed as single-channel signal. On the other hand, the signal associated with multi-variables is known as multi-channel signal. Mathematically, if signal $x(t)$ has $M$ channels, then it can be represented as,

$$x(t) = [x_1(t), x_2(t), \ldots, x_M(t)]^T \tag{1.10}$$

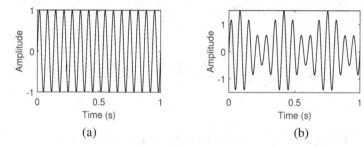

**Figure 1.9**    Example of (a) stationary signal and (b) non-stationary signal.

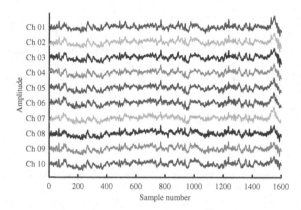

**Figure 1.10**    Example of multi-channel signal. Ch $i$ represents signal corresponding to channel $i$, where $i$ varies from 1 to 10.

The signals $x_1(t), x_2(t), x_3(t), \ldots, x_M(t)$ are known as component signals or channel signals. In biomedical engineering, such multi-channel signals are common such as, EEG, and ECG recordings from various electrodes at a time. Figure 1.10 shows an example of recorded multi-channel EEG signals for motor imagery task. These signals are obtained from the Physionet EEG database (EEG motor movement/imagery dataset) [3, 7].

### 1.2.8   ENERGY AND POWER SIGNALS

The energy signals have the finite energy. Energy of the signal $y(t)$ can be defined mathematically as [8],

$$E = \int_{-\infty}^{\infty} |y(t)|^2 dt \tag{1.11}$$

On the other hand, if the energy of a signal is infinite and the mean power is finite then signal $y(t)$ can be termed as power signal. The mean power of the signal $y(t)$ is expressed as,

$$P = \lim_{T \to \infty} \frac{1}{T} \int_{-T/2}^{T/2} |y(t)|^2 dt \tag{1.12}$$

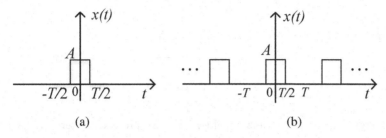

(a)                                                    (b)

**Figure 1.11**   Example of (a) energy signal and (b) power signal.

The periodic signals exist for infinite time and in order to generate such signals an infinite energy will be needed as continuous work is required. Such signals will have infinite energy but finite power, which are the example of power signals. One example of power signal is shown in Fig. 1.11 (b). On other side, time-limited signals with finite values like pulse will require finite energy to generate. A signal that falls in the category of energy signals, is shown in Fig. 1.11 (a).

## 1.3   VARIOUS MEASURES OF THE SIGNALS

### 1.3.1   NORMED SPACE

In normed space, the signals are considered like vectors which are elements of linear vector space denoted by $Y$. The norm of vector or signal $y$ can be considered as a length, size or strength of the vector or signal [9]. The norm is denoted by $||y||$.

The norm of the continuous-time signal $y(t)$ over the interval $t_1 \leq t \leq t_2$ can be represented as follows:

$$||y||_p = \left[ \int_{t_1}^{t_2} |y(t)|^p dt \right]^{1/p} ; \quad 1 \leq p \leq \infty \tag{1.13}$$

Here, $p = 2$ provides Euclidian-norm and can be written as,

$$||y||_2 = \sqrt{\int_{t_1}^{t_2} |y(t)|^2 dt} \; ; \quad y \in L_2(t_1, t_2) \tag{1.14}$$

The squaring operation of this norm provides energy of the signal and it can be expressed as,

$$E = \int_{-\infty}^{\infty} |y(t)|^2 dt = ||y||_2^2 ; \quad y \in L_2(R) \tag{1.15}$$

In case of discrete-time signal, $y(n)$ and it is defined over the range $N_1 \leq n \leq N_2$. These norms can be expressed as follows:

$$||y||_p = \left[ \sum_{n=N_1}^{N_2} |y(n)|^p \right]^{1/p} ; \quad 1 \leq p \leq \infty \tag{1.16}$$

The value $p = 2$ provides the following expression:

$$||y||_2 = \sqrt{\sum_{n=N_1}^{N_2} |y(n)|^2} \tag{1.17}$$

The energy of this discrete-time signal $y(n)$ can be represented in terms of the above-mentioned $l_2$ norm as,

$$E = \sum_{n=-\infty}^{\infty} |y(n)|^2 = ||y||_2^2; \quad y \in l_2(-\infty,\infty); \ n \in \mathbb{Z} \tag{1.18}$$

The discrete counter part of $L_p$ is denoted by $l_p$. As $p \to \infty$, for continuous-time signal, $||y||_\infty = \max_{t_1 \leq t \leq t_2} |y(t)|$ and for discrete-time signal,

$$||y||_\infty = \max_{0 \leq n \leq N-1} |y(n)|.$$

## 1.3.2 INNER PRODUCT SPACE

The most commonly used signal spaces in signal processing are $L_2$ (continuous-time signal) and $l_2$ (discrete-time signal). The inner product can be defined for these spaces. The notation $< y_1, y_2 >$ represents the process of calculating a number from the pair of vectors or signals. This operation is called inner product or dot product. The $L_2$ norm can also be defined in term of inner product which is represented as,

$$||y_1||_2 = \sqrt{<y_1,y_1>} \tag{1.19}$$

The inner product can be defined in following ways for different kind of signals in the Table 1.2.

### 1.3.2.1 Orthogonality Condition

Two vectors or signals are considered as orthogonal if the inner product between these vectors is zero, i.e.,

$$< y_1, y_2 >= 0 \tag{1.20}$$

## 1.3.3 METRIC SPACE

A metric space can be defined by $(Y,d)$, where set $Y$ and function $d$ provide a real number to each pair of elements $A$ and $B$ of $Y$, $d(A,B) \geq 0$. The metric $d(A,B)$ has

## Table 1.2

## Formulas for Inner Product Computation

| Type of Signals | Inner Product |
|---|---|
| Continuous-time energy signals | $<y_1,y_2> = \int_{-\infty}^{\infty} y_1(t)y_2^*(t)dt$ |
| Continuous-time power signals | $<y_1,y_2> = \lim_{T\to\infty} \frac{1}{2T} \int_{-T}^{T} y_1(t)y_2^*(t)dt$ |
| Discrete-time energy signals | $<y_1,y_2> = \sum_{n=-\infty}^{\infty} y_1(n)y_2^*(n)$ |
| Discrete-time power signals | $<y_1,y_2> = \lim_{N\to\infty} \frac{1}{2N+1} \sum_{n=-N}^{N} y_1(n)y_2^*(n)$ |

the following properties:

Property 1: $d(A,B) = 0$ if and only if $A = B$.

Property 2: $d(A,B) = d(B,A)$ (symmetric nature).

Property 3: $d(A,B) \leq d(A,C) + d(C,B)$ (triangle inequality concept).

The metric $d(A,B)$ can be seen as a difference measure between $B$ and $A$, and $d$ is termed as a metric or distance function. This measure is used to determine the distance between two vectors.

There are many ways to determine the distance between two sets, signals, and vectors. For example, distance between two vectors $X = [x_1,x_2,x_3]^T$ and $Y = [y_1,y_2,y_3]^T$ can be given by,

$$d_p(x,y) = [|x_1 - y_1|^P + |x_2 - y_2|^P + |x_3 - y_3|^P]^{1/p} \qquad (1.21)$$

where $1 \leq p \leq \infty$.

We can have various distances based on the $p$ values. For example, $p = 1$ provides,

$$d_1(x,y) = [|x_1 - y_1| + |x_2 - y_2| + |x_3 - y_3|]$$

For $p = 2$, Eq. (1.21) can be written as,

$$d_2(x,y) = \sqrt{[|x_1 - y_1|^2 + |x_2 - y_2|^2 + |x_3 - y_3|^2]}$$

For continuous-time signals $y_1(t)$ and $y_2(t)$ over interval $t_1 \leq t \leq t_2$, these distance measures are given by,

$$d_p(y_1,y_2) = \left[\int_{t_1}^{t_2} |y_1(t) - y_2(t)|^P dt\right]^{1/p}, \quad y_1 \text{ and } y_2 \in L_2(t_1,t_2) \qquad (1.22)$$

For $p = 2$ and $\infty$,

$$d_2(y_1, y_2) = \sqrt{\int_{t_1}^{t_2} |y_1(t) - y_2(t)|^2 dt} \tag{1.23}$$

$$d_\infty(y_1, y_2) = \max_{t_1 \le t \le t_2} |y_1(t) - y_2(t)| \tag{1.24}$$

These distance measures for discrete-time signals can be defined as,

$$d_p(y_1, y_2) = \left[ \sum_{n=0}^{N-1} |y_1(n) - y_2(n)|^p \right]^{1/p} , \quad y_1 \text{ and } y_2 \in l_2(0, N-1) \tag{1.25}$$

For $p = 2$ and $\infty$:

$$d_2(y_1, y_2) = \sqrt{\sum_{n=0}^{N-1} |y_1(n) - y_2(n)|^2} \tag{1.26}$$

and

$$d_\infty(y_1, y_2) = \max_{0 \le n \le N-1} |y_1(n) - y_2(n)| \tag{1.27}$$

There are some other matrices which are not related by norm as described above, for example, Hamming distance which finds application in coding theory.

## 1.4   IMPORTANT SIGNALS

### 1.4.1   SINUSOIDAL SIGNALS

These signals are commonly used for simple form representation of the signals. Mathematically, these signals can be written as,

$$y(t) = A\cos(\omega t + \phi) \tag{1.28}$$

Where $A$ is termed as amplitude, $\omega$ is known as frequency, and $\phi$ denotes phase of the sinusoidal form of the signal. The relation between frequency $\omega$ in radian per second and frequency $f$ in Hz is given by,

$$\omega = 2\pi f = \frac{2\pi}{T} \text{ with } f = \frac{1}{T} \tag{1.29}$$

Graphical plot of the such signal is shown in Fig. 1.12.

### 1.4.2   UNIT-STEP FUNCTION

The continuous-time unit-step function is a basic test signal and can be defined as,

$$u(t) = \begin{cases} 1, & t > 0 \\ 0, & t < 0 \end{cases} \tag{1.30}$$

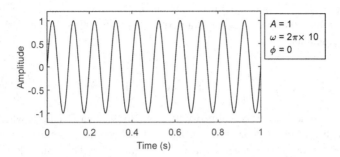

**Figure 1.12**  Plot of sinusoidal signal.

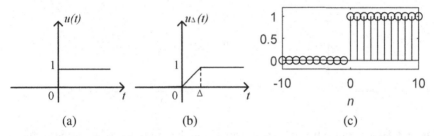

(a)                         (b)                         (c)

**Figure 1.13** (a) Continuous-time unit-step function, (b) representation of continuous-time unit-step function in limiting way, and (c) plot of discrete unit-step function.

and its plot is shown in Fig. 1.13 (a). The unit-step function $u(t)$ is discontinuous at $t = 0$ and more practically, it is defined in limiting way ($\lim_{\Delta \to 0} u_\Delta(t) = u(t)$) as shown in Fig. 1.13 (b).

The discrete version of the unit-step function is expressed as,

$$u(n) = \begin{cases} 1, & n \geq 0 \\ 0, & n < 0 \end{cases} \tag{1.31}$$

and plot of discrete-time unit-step function $u(n)$ can be seen in Fig. 1.13 (c).

### 1.4.3 IMPULSE FUNCTION

It is also an important function in signal processing. The impulse function $\delta(t)$, which is also known as Dirac delta function, has the following two properties:

$$\delta(t) = \begin{cases} 0, & t \neq 0 \\ \text{not defined}, & t = 0 \end{cases} \tag{1.32}$$

$$\int_{-\infty}^{\infty} \delta(t)dt = 1 \tag{1.33}$$

The impulse function is defined with the help of other function with the limiting process. It can also be expressed as a derivative of the unit-step function (ideal and limiting cases) as,

$$\text{Ideal case:} \quad \delta(t) = \frac{du(t)}{dt}, \quad \text{Limiting case:} \lim_{\Delta \to 0} \delta_\Delta(t) = \lim_{\Delta \to 0} \frac{du_\Delta(t)}{dt} \quad (1.34)$$

Impulse function can be defined with the help of other functions which has unit area and in limiting approach, functions tend to become an impulse. Such signals include, Gaussian, sinc, wavelet, pulse, etc. Figure 1.14 shows an impulse function generating from the limiting process of rectangular pulse function whose area is unity and pulse width approaches to zero with the limiting process.

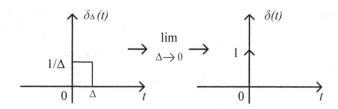

**Figure 1.14**   Impulse function generation from the limiting process of pulse function.

The unit-step function can be obtained by the use of running integral over impulse function, i.e.,

$$u(t) = \int_{-\infty}^{t} \delta(\tau)d\tau \quad (1.35)$$

The sampling property of an impulse is given by,

$$p(t)\delta(t - \tau) = p(\tau)\delta(t - \tau) \quad (1.36)$$

The impulse function $\delta(t)$ is an even function, i.e., $\delta(t) = \delta(-t)$ and has the following sifting property:

$$\int_{-\infty}^{\infty} p(t)\delta(t - \tau)dt = p(\tau) \quad (1.37)$$

### 1.4.4   RAMP FUNCTION

In case of unit-step function it changes its state from 0 to 1 instantaneously, whereas in a ramp function state changes linearly. The ramp function is shown in Fig. 1.15.
Mathematically, it can be expressed as,

$$r(t) = \frac{1}{\Delta}t, \quad \text{for } 0 \leq t; \ r(t) = 0, \text{ for } 0 > t \quad (1.38)$$

The ramp function $r(t)$ can be expressed in terms of unit-step function $u(t)$ as, $r(t) = \frac{1}{\Delta}tu(t)$.

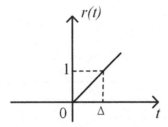

**Figure 1.15**  Ramp function.

## 1.5  SIGNAL OPERATIONS

The commonly used signal operations include time shifting, time scaling, and time reversal [10]. These operations are explained below.

### 1.5.1  TIME SHIFTING

It is performed by shifting the independent variable of the signal. For example, $x(t - T)$ shifts the $x(t)$ to right by $T$ amount, termed as delay, and similarly $x(t + T)$ shifts the $x(t)$ to left by $T$ amount, known as advance operation. Figure 1.16 shows the delay operation for a signal by an amount of $T$. In the plot of shifted signal $x(t - T)$, the actual signal is shown using dashed line.

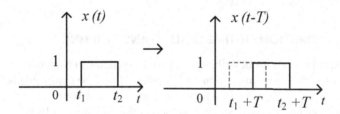

**Figure 1.16**  Delay of the signal by $T$ amount.

### 1.5.2  TIME SCALING

This operation is required to compress and dilate the signals. The scaled signal is represented by $x(at)$. If $a > 1$, then signal becomes compressed. When $a < 1$ then, signal becomes dilated or expanded. The dilation and compression operations for a signal are demonstrated in Fig. 1.17. In the plots of scaled signals, the actual signal has been shown using dashed line.

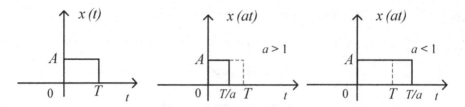

**Figure 1.17**   Scaling operation on a signal $x(t)$.

### 1.5.3   TIME REVERSAL

This operation on the signal results in flipping of the signal with respect to y-axis. It is performed by replacing the independent variable $t$ by $-t$. Figure 1.18 shows the time reversal process for a signal $x(t)$.

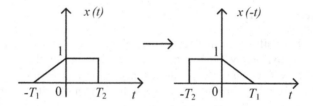

**Figure 1.18**   Time reversal operation.

### 1.5.4   COMBINATION OF TIME SCALING AND SHIFTING

Three different operations described in the above section can be combined together. The order of performing time scaling or reversing should be completed before time shifting to get $x\left(\frac{t-T}{a}\right)$ from $x(t)$, where $T$ is amount of shifting and $a$ is time scaling factor which can take negative values and for that time scaling involves time reversal. Both the scaling and shifting operations on a signal $x(t)$ are shown in following mathematical expression:

$$x(t) \rightarrow x\left(\frac{t-T}{a}\right) \tag{1.39}$$

### PROBLEMS

Q 1.1   Determine area of each of the following functions and suggest modification to make unit area so that impulse function can be obtained by using limiting process of these modified functions. In each case, show the limiting process in order to obtain impulse function.

(a) Gaussian function $x(t) = e^{-\frac{t^2}{2\sigma^2}}$

(b) Gate function $x(t) = 1$ for $-T \le t \le T$

(c) Sinc function $x(t) = \text{sinc}\left(\frac{t}{2\tau}\right)$, where $\text{sinc}(t) = \frac{\sin(\pi t)}{\pi t}$

(d) Triangular function $\begin{cases} \frac{t}{2\Delta} + 1 & ; -2\Delta \le t \le 0 \\ \frac{-t}{2\Delta} + 1 & ; 0 \le t \le 2\Delta \\ 0 & ; \text{otherwise} \end{cases}$

(e) $e^{-\beta t} \, u(t)$, for $\beta > 0$

Q 1.2 Suppose $g(t) = \begin{cases} 0 & ; \ t < 0 \\ t^2 & ; \ t \ge 0 \end{cases}$. Sketch the following:

(a) $g(t)$

(b) $g(t-2)$

(c) $g(t+2)$

(d) $g(-t)$

Q 1.3 Suppose $x(t) = \begin{cases} 0; & t < 0 \\ 2t; & 0 \le t \le 1 \\ 0; & t > 1 \end{cases}$. Sketch the following:

(a) $x(t)$

(b) $x(t/4)$

(c) $x(4t)$

(d) $x(t/2)$

(e) $x(2t)$

Q 1.4 Explain why the following signals are non-stationary:

(a) Amplitude modulated signal: $x(t) = A[1 + \mu \, g(t)]\cos(2\pi f_c t + \theta)$, where $g(t)$ is the message, $\mu$ is the modulation index, $f_c$ is the carrier frequency, $\theta$ is an arbitrary phase angle, and $A$ is carrier amplitude.

(b) Frequency modulated signal: $x(t) = A\cos\left[2\pi f_c t + \mu \int_{t_s}^{t} g(\tau) d\tau + \theta\right]$, where $g(t)$ is message, $A$ is carrier amplitude, $f_c$ is the carrier frequency, $\theta$ is an arbitrary angle, $\mu$ is modulation index, and $t_s$ represents the starting time for frequency modulated transmission.

(c) Damped cosine function: $x(t) = e^{-\sigma t}\cos(\omega_0 t)$, where $\sigma$ is the damping factor and $\omega_0$ is the angular frequency.

(d) Bessel function: $x(t) = J_0\left(\frac{\lambda_m}{a}t\right)$, where $\lambda_m$ is the $m^{\text{th}}$ root of Bessel function (zero-order first kind) $J_0(\cdot) = 0$ and $a$ represents the duration of the signal.

(e) Gaussian modulated signal (impulse response of Gabor filter): $x(t) = e^{-\sigma t^2}\cos(\omega_0 t)$, where $\sigma$ controls the bandwidth of the signal and $\omega_0$ denotes the carrier frequency.

Q 1.5 Simplify the following expression: $x(t) = t^4\delta(t-1) + \sin(2t)\delta(t) + \cos(2t)\delta(t) + t^4\delta(t+1)$. Sketch the final output.

Q 1.6 For the gate function,

$$x(t) = \begin{cases} 1, & -1 \le t \le 0 \\ 0, & \text{otherwise} \end{cases}$$

Determine and sketch the following signals:

(a) $x(t-4)$

(b) $x(t+4)$

(c) $x(\frac{3}{2}t)$

(d) $x(\frac{2}{3}t)$

(e) $x(-t+4)$

(f) $x(-t-4)$

(g) $x(2t-3)$

Q 1.7 Determine whether the following signals are even or odd signals.

(a) $e^{-5t}$

(b) $u(t+10)$

(c) $e^{-\sigma t^2}$

(d) $e^{-\sigma t^2}\sin(\omega_0 t)$

(e) $\text{sinc}(t) = \frac{\sin(\pi t)}{\pi t}$

(f) $\delta(t)$

Q 1.8 Determine whether the following signals are energy or power signals.

(a) $e^{\sigma t}u(t)$, where $\sigma > 0$

(b) $A\cos(\omega_0 t)[u(t) - u(t-T)]$, where $\omega_0 = 2\pi/T$

(c) $A\cos(\omega_0 t)u(t)$

(d) $t^n u(t)$, where $n \ge 1$

(e) $t^n u(t)$, where $n = 0$

Q 1.9 Determine whether the following signals are periodic or aperiodic signals.

(a) $5 + \cos(4\pi t)$

(b) $e^{-5|t|}$

(c) $x(t) = x_1(t) + x_2(t) + x_3(t) + \cdots + x_M(t)$. Given that $x_1(t), x_2(t), x_3(t), \ldots,$ $x_M(t)$ are periodic signals with finite periods of $T_1, T_2, T_3, \ldots, T_M$, respectively.

    (d) $5 + \cos(4\pi t)u(t)$

    (e) $5 + \cos(4\pi t) + 2\delta(t)$

    Also, determine the period.

Q 1.10 Determine the energy of the following signals:

    (a) Modulated signal: $\phi(t) = g(t)e^{j\omega_0 t}$, here $g(t)$ is a message signal with energy of $E_g$ and $\omega_0$ denotes the carrier frequency.

    (b) Daughter wavelet: $\psi_{a,b}(t) = \frac{1}{\sqrt{a}}\psi\left(\frac{t-b}{a}\right)$, here $\psi(t)$ is mother wavelet with energy of $E_\psi$, $a$ denotes the scale parameter, and $b$ is the shifting parameter.

Q 1.11 Suppose signal $x(t) = 3t + 2$ for $0 \leq t \leq 2$ determine norm 1, norm 2, norm 3, and norm $\infty$. Also, determine the energy of the signal based on norm 2.

Q 1.12 Determine the inner product, $L_2$ norm of each signal, and distance between both of them for the following two signals: $x_1(t) = e^{j3\omega_0 t}$ and $x_2(t) = e^{j6\omega_0 t}$, where $\omega_0 = \frac{2\pi}{T}$. Consider distance of two signals as squared root of inner product of the difference of two signals $x_1(t) - x_2(t)$ with itself, when the signals are power signals.

# 2 Signal Representation

*"A dream is not that which you see while sleeping, it is something that does not let you sleep."*– APJ Abdul Kalam

## 2.1 SIGNAL REPRESENTATION IN TERMS OF ORTHOGONAL FUNCTIONS

In real life, signals are complicated in nature and these signals do not provide proper analysis. In order to have better analysis, it is useful to represent these complicated signals in terms of simple signals or basis functions or basic signals or known signals. Such kind of representation also helps to represent complicated or unknown signals in terms of simple signals or known signals. Analysis of signal in terms of basis functions becomes easier. It is also preferred to have orthogonality in the basis functions so that representation of the signals will be unique.

## 2.2 SIGNAL REPRESENTATION IN TERMS OF IMPULSE FUNCTIONS

The sampling theorem provides a bridge between analog signals (or continuous-time signals) and discrete-time signals [11]. Based on this theorem, a minimum sampling rate can be decided by which analog signals can be uniformly sampled so that original analog signal can be reconstructed or recovered from these samples. The minimum sampling rate, which is termed as Nyquist rate, is twice of the highest frequency present in the signal. The sampling process of the continuous-time signals provides a set of samples, and these samples can be represented in terms of impulses as a basis function set. For example, a discrete-time signal $x(n)$, $-\infty < n < \infty$ can be represented in terms of impulses as follows:

$$x(n) = \sum_{k=-\infty}^{\infty} x(k)\,\delta(n-k) \tag{2.1}$$

where $\delta(n-k)$ is defined as,

$$\delta(n-k) = \begin{cases} 1 & n=k \\ 0 & n \neq k \end{cases} \tag{2.2}$$

In this representation, $x(k)$ are termed as coefficients and $\delta(n-k)$ that is known as Kronecker delta function serves as basis function [12]. We can see that this representation provides the same number of coefficients as number of samples in the signal. Due to this reason, impulse function-based representation is not suitable for compression of the signal based on the fewer coefficients. Moreover, the basis function does not provide any characterization, so these coefficients are not suitable for feature representation.

DOI: 10.1201/9781003367987-2

Equation (2.1) for finite interval ($0 \leq n \leq N-1$) can be written as follows:

$$x(n) = \sum_{k=0}^{N-1} x(k)\, \delta(n-k) \tag{2.3}$$

Define vector $\bar{x}$ as follows:

$$\bar{x} = [x(0), x(1), x(2), \ldots, x(N-1)]^T \tag{2.4}$$

$$= x(0)[1,0,0,\ldots,0]^T + x(1)[0,1,0,\ldots,0]^T + \cdots + x(N-1)[0,0,0,\ldots,1]^T \tag{2.5}$$

Now the vector $\bar{x}$ can be written as,

$$\bar{x} = x(0)\bar{e}_0 + x(1)\bar{e}_1 + \cdots + x(N-1)\bar{e}_{N-1} \tag{2.6}$$

where $\bar{e}_0 = [1,0,0,\ldots,0]^T$, $\bar{e}_1 = [0,1,0,\ldots,0]^T$, and $\bar{e}_{N-1} = [0,0,0,\ldots,1]^T$. These basis functions in Eq. (2.6) are orthogonal to each other, i.e.,

$$\sum_{n=0}^{N-1} e_i(n) e_j(n) = \bar{e}_i^T \bar{e}_j = 0 \quad \text{for} \quad i \neq j \tag{2.7}$$

It should be noted that in this representation, norm of each basis function is unity or equal to one, i.e.,

$$\sum_{n=0}^{N-1} |e_i(n)|^2 = 1, \quad \text{for} \quad i = 0,1,2,\ldots,N-1. \tag{2.8}$$

Due to the unity norm of these basis functions, they are also part of orthonormal basis functions families.

## 2.3  SIGNAL REPRESENTATION IN TERMS OF GENERAL BASIS FUNCTIONS

Consider a general basis function $\phi_m(n)$, then,

$$\sum_{n=0}^{N-1} \phi_m^*(n)\phi_l(n) = a_m^2 \delta(m-l)$$

$$= \begin{cases} a_m^2, & \text{for } m = l \\ 0, & \text{otherwise} \end{cases} \tag{2.9}$$

where $a_m$ is the norm of $\phi_m(n)$. The asterisk denotes the complex conjugate operation. The orthonormal basis functions can be obtained by proper normalization of the orthogonal basis functions as expressed in the equation below,

$$\phi_{\text{norm},m}(n) = \frac{\phi_m(n)}{a_m}, \quad 0 \leq m \leq N-1 \tag{2.10}$$

Based on these basis functions, any given signal can be represented and corresponding analysis and synthesis equations are as follows:

$$\text{Synthesis equation:} \quad x(n) = \sum_{k=0}^{N-1} \theta(k)\phi_{\text{norm},k}(n), \quad 0 \leq n \leq N-1 \qquad (2.11)$$

$$\text{Analysis equation:} \quad \theta(k) = \sum_{n=0}^{N-1} x(n)\phi_{\text{norm},k}^*(n), \quad 0 \leq k \leq N-1 \qquad (2.12)$$

This signal representation is also termed as generalized Fourier representation and $\theta(k)$ denotes the generalized Fourier coefficients. Equation (2.12) can be proved with the help of Eq. (2.11) by multiplying both sides $\phi_{\text{norm},k}^*(n)$ and performing summation over the index $n$ and applying orthogonality property of basis functions.

Parseval's theorem for this general form of representation can be stated as follows:

$$\sum_{n=0}^{N-1} |x(n)|^2 = \sum_{k=0}^{N-1} |\theta(k)|^2 \qquad (2.13)$$

Equation (2.13) can be proved with the help of Eq. (2.11), by multiplying signal $x^*(n)$ and summing over $n$. Based on this theorem, the energy of the signal based on the samples is similar to the energy based on the generalized Fourier spectral coefficients.

In Eq. (2.12), the spectral coefficients $\theta(k)$ can provide an approximation of $x(n)$ in least square way. Suppose, the signal $x(n)$ is approximated by the $N_1$ basis functions out of total $N$ basis functions. In this case, $\hat{\theta}(k)$, $k = 0,1,2,...,N_1 - 1$ will be the least square solution. The same can be proved as follows:
The approximation signal can be expressed as follows:

$$\hat{x}(n) = \sum_{k=0}^{N_1-1} \hat{\theta}(k)\phi_{\text{norm},k}(n) \qquad (2.14)$$

Whereas the actual signal can be written as,

$$x(n) = \sum_{k=0}^{N-1} \theta(k)\phi_{\text{norm},k}(n) \qquad (2.15)$$

and, corresponding approximation error can be written as,

$$e(n) = x(n) - \hat{x}(n) \qquad (2.16)$$

The coefficients $\hat{\theta}(k)$ need to be such that the square error is minimized. The error energy is computed as follows:

$$\begin{aligned} E_{N_1} &= \sum_{n=0}^{N-1} |e(n)|^2 = \sum_{n=0}^{N-1} |x(n) - \hat{x}(n)|^2 \\ &= \sum_{n=0}^{N-1} |x(n)|^2 + \sum_{n=0}^{N-1} |\hat{x}(n)|^2 - 2\sum_{n=0}^{N-1} x(n)\hat{x}(n) \end{aligned}$$

By using Parseval's relation, the error energy can be written as,

$$E_{N_1} = \sum_{k=0}^{N-1} |\theta(k)|^2 + \sum_{k=0}^{N_1-1} |\hat{\theta}(k)|^2 - 2 \sum_{k=0}^{N_1-1} \theta(k)\hat{\theta}(k)$$

This error energy expression can be written as,

$$E_{N_1} = \sum_{k=0}^{N_1-1} |\theta(k)|^2 + \sum_{k=N_1}^{N-1} |\theta(k)|^2 + \sum_{k=0}^{N_1-1} |\hat{\theta}(k)|^2 - 2 \sum_{k=0}^{N_1-1} \theta(k)\hat{\theta}(k)$$

The above expression can be further simplified as,

$$E_{N_1} = \sum_{k=N_1}^{N-1} |\theta(k)|^2 + \sum_{k=0}^{N_1-1} |\theta(k) - \hat{\theta}(k)|^2$$

It should be noted that the first term is independent of $\hat{x}(n)$ or $\hat{\theta}(k)$. Therefore, the error energy $E_{N_1}$ will be minimized only when $\theta(k) = \hat{\theta}(k)$.

It can also be proved in the following way:

$$\frac{\partial E_{N_1}}{\partial \hat{\theta}} = 0, \quad \theta(k) = \hat{\theta}(k)$$

When $N_1 = N$, then $E_{N_1} = 0$. Similarly, it can be shown that $E_{N_1+1} \leq E_{N_1}$ for any $N_1$.

It should be noted that when we use $N_1$ basis functions out of $N$ basis functions in the approximation of the signal, total basis function space $\{\phi_0(n), \phi_1(n), \phi_2(n), \ldots, \phi_{N_1-1}(n), \phi_{N_1}(n), \ldots, \phi_{N-1}(n)\}$ cover the approximation and error vector spaces.

Approximation estimate $\hat{x}(n)$ and error $e(n)$ cover the following vector spaces:

$$\text{Approximation space:} \quad V_1 = \{\phi_0(n), \phi_1(n), \ldots, \phi_{N_1-1}(n)\}$$
$$\text{Error space:} \quad V_2 = \{\phi_{N_1}(n), \phi_{N_1+1}(n), \ldots, \phi_{N-1}(n)\}$$

It can be seen that every vector in $V_1$ is orthogonal to every vector in $V_2$. The vector space $V_1$ is orthogonal to the vector space $V_2$. This gives the concept of data and error orthogonality, i.e.,

$$\sum_{n=0}^{N-1} e(n)\hat{x}(n) = 0 \tag{2.17}$$

The $e(n)$ and $\hat{x}(n)$ can be expressed as follows:

$$e(n) = \sum_{k=N_1}^{N-1} \theta(k)\phi_k(n)$$

$$\hat{x}(n) = \sum_{k=0}^{N_1-1} \theta(k)\phi_k(n)$$

and with orthogonality property, Eq. (2.17) can be proved.

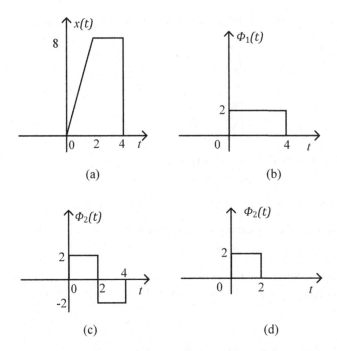

**Figure 2.1** (a) Signal under analysis, (b) first basis function, (c) second basis function, and (d) second-modified basis function.

When the basis functions are part of orthogonal functions, in that case, spectral coefficients hold the property of the finality of coefficients [13]. But when basis functions are non-orthogonal, then finality of coefficients does not hold. Here, the concept of fixed nature of coefficients (i.e., finality of the coefficients) will be demonstrated with the help of an example.

Suppose, we want to approximate signal $x(t)$, shown in Fig. 2.1 (a) with the help of basis functions $\phi_1(t)$ and $\phi_2(t)$ in interval $0 \leq t \leq 4$ shown in Figs. 2.1 (b) and (c), respectively. In the first case, we use only one basis function $\phi_1(t)$ in the approximation, and for the second case, we use both basis functions for the approximation. In both cases, we need to determine the coefficients and also check the property of fixed nature of coefficients.

Case 1: Basis function $\phi_1(t)$ shown in Fig. 2.1 (b) is used for approximation. The approximated signal in this case is given by,

$$\hat{x}(t) = c_1 \phi_1(t)$$

The error energy corresponding to this approximation signal can be given as,

$$
\begin{aligned}
E_1 &= \int_0^4 [x(t) - \hat{x}(t)]^2 \, dt \\
&= \int_0^2 (4t - 2c_1)^2 dt + \int_2^4 (8 - 2c_1)^2 dt \\
&= 16 \left[ \frac{32}{3} + c_1^2 - 6c_1 \right]
\end{aligned}
\tag{2.18}
$$

In order to minimize the error energy,

$$
\frac{dE_1}{dc_1} = 0 \text{ provides } c_1 = 3
$$

and corresponding error energy can be computed using Eq. (2.18), which gives $E_1 = \frac{80}{3}$.

Case 2: Basis functions $\phi_1(t)$ and $\phi_2(t)$, shown in Figs. 2.1 (b) and (c), respectively are used for approximation. The approximated signal can be given as,

$$
\hat{x}(t) = c_1 \phi_1(t) + c_2 \phi_2(t)
$$

and corresponding error energy is computed as,

$$
\begin{aligned}
E_2 &= \int_0^4 [x(t) - \hat{x}(t)]^2 \, dt \\
&= \int_0^4 [x(t) - c_1 \phi_1(t) - c_2 \phi_2(t)]^2 \, dt \\
&= \int_0^2 (4t - 2c_1 - 2c_2)^2 dt + \int_2^4 (8 - 2c_1 + 2c_2)^2 dt \\
&= 16 \left[ \frac{32}{3} + c_1^2 + c_2^2 - 6c_1 + 2c_2 \right]
\end{aligned}
\tag{2.19}
$$

In order to minimize the error,

$$
\frac{\partial E_2}{\partial c_1} = 0 \text{ provides } c_1 = 3, \text{ and } \frac{\partial E_2}{\partial c_2} = 0 \text{ provides } c_2 = -1
$$

The corresponding error energy can be computed using Eq. (2.19), which gives $E_2 = \frac{32}{3}$.

$$
E_2 < E_1, \quad \frac{32}{3} < \frac{80}{3}
$$

Case 3: Now, in order to study the non-orthogonal basis functions case, $\phi_2(t)$ is modified as shown in Fig 2.1 (d), so that the basis functions $\phi_1(t)$ and $\phi_2(t)$ are not orthogonal. $\phi_1(t)$ is considered the same, which is shown in Fig. 2.1 (b).

Now, in this situation, the error energy is given by,

$$E_2 = \int_0^4 [x(t) - \hat{x}(t)]^2 \, dt$$

$$= \int_0^4 [x(t) - c_1\phi_1(t) - c_2\phi_2(t)]^2 \, dt$$

$$= \int_0^2 (4t - 2c_1 - 2c_2)^2 dt + \int_2^4 (8 - 2c_1)^2 dt$$

$$= 4\left(\frac{128}{3} + 4c_1^2 + 2c_2^2 - 24c_1 - 8c_2 + 4c_1c_2\right) \qquad (2.20)$$

In order to minimize the error energy,

$$\frac{\partial E_2}{\partial c_1} = 0 \text{ gives } 2c_1 + c_2 = 6, \text{ and } \quad \frac{\partial E_2}{\partial c_2} = 0 \text{ gives } c_1 + c_2 = 2$$

From the above equations, we have, $c_1 = 4$, and $c_2 = -2$. The error energy in this case is computed using Eq. (2.20), which is $E_2 = \frac{32}{3}$.

The coefficients obtained from Case 3 show that non-orthogonal basis functions do not follow fixed nature of the coefficients condition. It should be noted that in this case,

$$E_2 < E_1, \quad \frac{32}{3} < \frac{80}{3}$$

i.e., the error is reduced but coefficients are not final or fixed. This example shows the need of orthogonality in basis functions in order to have unique representation.

## 2.4  SIGNAL REPRESENTATION IN TERMS OF COMPLEX EXPONENTIAL FUNCTIONS

The complex exponential functions (sinusoidal signals) are very useful in signal processing [14, 15]. The representation of the signal in terms of these basis functions provides frequency-domain for signal analysis. The complex exponential functions when they are given as an input to a linear time invariant (LTI) system, then output of such system also provides complex exponential functions, which are also termed as eigenfunctions of the LTI system. Another reason for using complex exponential functions (sinusoidal functions) in signal processing is easy and convenient nature of such signals for analysis.

### 2.4.1  FOURIER SERIES REPRESENTATION

A periodic signal can be represented based on orthogonal and harmonically related complex exponential functions [6]. The conditions, which are known as Dirichlet's conditions, need to be satisfied by the signal in order to have Fourier series representation.

These conditions include single-valued nature of the function, finite number of discontinuities, finite number of maxima and minima, and absolutely integrable nature of the signal in the interval in which we want to analyze the signal. Trigonometric form of Fourier series is given as follows in the interval $(0, T)$:

$$\text{Synthesis equation:} \quad x(t) = c_0 + \sum_{m=1}^{\infty} c_m \cos(m\omega_0 t) + d_m \sin(m\omega_0 t) \quad (2.21)$$

The coefficients $c_0$, $c_m$, and $d_m$ are computed as follows:

$$\text{Analysis equation:} \quad \begin{cases} c_0 = \frac{1}{T} \int_0^T x(t)\, dt \\ c_m = \frac{2}{T} \int_0^T x(t) \cos(m\omega_0 t)\, dt \\ d_m = \frac{2}{T} \int_0^T x(t) \sin(m\omega_0 t)\, dt \end{cases} \quad (2.22)$$

Equation (2.21) can be written in the following form:

$$x(t) = a_0 + \sum_{m=1}^{\infty} a_m \cos(m\omega_0 t + \phi_m) \quad (2.23)$$

Where,

$$a_0 = c_0$$
$$a_m = \sqrt{c_m^2 + d_m^2}$$
$$\phi_m = \tan^{-1} \frac{-d_m}{c_m} \qquad \text{and} \quad T = \frac{2\pi}{\omega_0}$$

The Fourier series can also be expressed in the complex exponential form as follows in the interval $(0, T)$:

$$\text{Synthesis equation:} \quad x(t) = \sum_{m=-\infty}^{\infty} g_m e^{jm\omega_0 t} \quad (2.24)$$

Where,

$$\text{Analysis equation:} \quad g_m = \frac{1}{T} \int_0^T x(t) e^{-jm\omega_0 t}\, dt \quad (2.25)$$

These two forms (2.21) and (2.24) can be obtained from each other. It should also be noted that the complex exponential form of the Fourier series contains negative frequency components. These frequency components do not convey any physical meaning. They exist due to the mathematical relation between cosine and complex exponential functions.

It should be noted that the signal under analysis using Fourier series expansion is periodic. The signal part repeats itself over every interval of time and Fourier series expansion coefficients are same for every interval, so any signal part can be used for computing the Fourier series coefficients. Hence, the Fourier series representations in Eqs. (2.21) and (2.24) are valid for any interval of period in the signal. For an example, the Fourier series coefficients for periodic pulse signal shown in Fig. 2.2 (a) can be computed as follows:

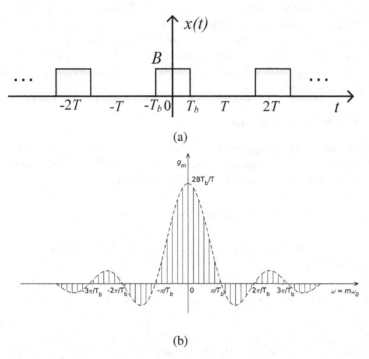

(a)

(b)

**Figure 2.2**  (a) A periodic rectangular pulse $x(t)$ and (b) Fourier series coefficients $g_m$ of the periodic rectangular pulse $x(t)$.

$$g_m = \frac{1}{T} \int_{-T_b}^{T_b} B e^{-jm\omega_0 t} dt$$

$$= \frac{2B}{m\omega_0 T} \sin(m\omega_0 T_b)$$

This expression of the Fourier coefficients can be used to compute $g_m$ for $m \neq 0$. For $m = 0$, the Fourier series coefficient is as follows:

$$g_0 = \frac{1}{T} \int_{-T_b}^{T_b} B\, dt = \frac{2BT_b}{T}$$

Substituting $\omega = m\omega_0$, the Fourier series coefficients can be expressed as,

$$g_m = \frac{2B}{\omega T} \sin(\omega T_b) = \frac{2BT_b}{T} \mathrm{Sa}(\omega T_b)$$

Here, $\text{Sa}(\cdot)$ denotes sampling function and $\text{Sa}(\omega T_b) = \frac{\sin(\omega T_b)}{\omega T_b}$. These coefficients can also be represented in terms of sinc function with $\text{sinc}(x) = \frac{\sin(\pi x)}{\pi x}$. Zero-crossing takes place when, $\omega T_b = \pm n\pi$, which gives $\omega = \frac{\pm n\pi}{T_b}$ [16].

The Fourier series coefficients $g_m$ are sketched in Fig. 2.2 (b). The spacing between two consecutive spectral lines is $\omega_0 = \frac{2\pi}{T}$. When $\omega_0 = 0$, it means $T \to \infty$, this provides an extension of Fourier series to Fourier transform. In real life, most of the signals are aperiodic and for such signals we can assume the time period as $\infty$. This means that signals will repeat after $\infty$ time, which also means that they will never repeat and for such signals the spectral lines shown in Fig. 2.2 (b) become a continuous spectrum.

### 2.4.2  FOURIER TRANSFORM

Fourier transform can be applied to continuous-time periodic and aperiodic signals in order to obtain their frequency contents or spectrum. The analysis and synthesis expressions for the Fourier transform are given as follows:

$$\text{Analysis expression: } X(\omega) = \int_{-\infty}^{\infty} x(t)e^{-j\omega t}dt \tag{2.26}$$

$$\text{Synthesis expression: } x(t) = \frac{1}{2\pi}\int_{-\infty}^{\infty} X(\omega)e^{j\omega t}d\omega \tag{2.27}$$

In the previous example, periodic pulse signal with period $T \to \infty$ can be considered as aperiodic signal with only one pulse. In that case, its Fourier transform can be given by,

$$X(\omega) = \int_{-T_b}^{T_b} Be^{-j\omega t}dt$$
$$= \frac{2B}{\omega}\sin(\omega T_b) = 2BT_b\text{Sa}(\omega T_b)$$

The pulse in time-domain and corresponding continuous spectrum are sketched in Fig. 2.3. In the figure, the zero-crossing occurs when $\omega T_b = \pm n\pi$.

The Fourier transform has some interesting properties, and these properties are listed in Table 2.1.

We have seen earlier, the signals can be classified into energy and power signals and based on the nature of the signal (continuous or discrete), these signals further can be classified as follows:

1. Continuous-time energy signals

2. Continuous-time power signals

3. Discrete-time energy signals

4. Discrete-time power signals

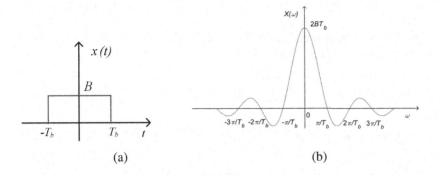

**Figure 2.3** (a) Rectangular pulse and (b) Fourier transform spectrum of the rectangular pulse.

**Table 2.1**
**Various Properties of the Fourier Transform**

| Property | Expressions |
|---|---|
| Linearity | $\sum\limits_{i=1}^{N} a_i x_i(t) \leftrightarrow \sum\limits_{i=1}^{N} a_i X_i(\omega)$ |
| Time-scaling | $x(at) \leftrightarrow \frac{1}{\|a\|} X\left(\frac{\omega}{a}\right)$ |
| Duality | $X(t) \leftrightarrow 2\pi x(-\omega)$ |
| Time-shifting | $x(t - \tau) \leftrightarrow e^{-j\omega\tau} X(\omega)$ |
| Frequency-shifting | $e^{j\omega_0 t} x(t) \leftrightarrow X(\omega - \omega_0)$ |
| Time differentiation | $\frac{dx(t)}{dt} \leftrightarrow (j\omega)X(\omega)$ and $\frac{d x^n(t)}{dt^n} \leftrightarrow (j\omega)^n X(\omega)$ |
| Time integration | $\int\limits_{-\infty}^{t} x(\tau)d\tau \leftrightarrow \frac{1}{j\omega}X(\omega) + \pi X(0)\delta(\omega)$ |
| Frequency differentiation | $-jtx(t) \leftrightarrow \frac{dX(\omega)}{d\omega}$ |
| Parseval's relation | $\int\limits_{-\infty}^{\infty} \|x(t)\|^2 dt = \frac{1}{2\pi} \int\limits_{-\infty}^{\infty} \|X(\omega)\|^2 d\omega$ |
| Time convolution or frequency multiplication theorem | $x_1(t) * x_2(t) \leftrightarrow X_1(\omega)X_2(\omega)$ |
| Frequency convolution or time multiplication theorem | $x_1(t)x_2(t) \leftrightarrow \frac{X_1(\omega)*X_2(\omega)}{2\pi}$ |

Fourier transform pairs are defined as, $x_1(t) \leftrightarrow X_1(\omega)$, $x_2(t) \leftrightarrow X_2(\omega)$, ..., $x_N(t) \leftrightarrow X_N(\omega)$. $*$ indicates the convolution operation.

For each class, we have corresponding Fourier transformation, which is shown in Table 2.2.

The discrete Fourier transform (DFT) based analysis and synthesis expressions using the concept of DTFS can be given follows:

$$\text{Analysis expression: } X(k) = \sum_{n=0}^{N-1} x(n)e^{-j2\pi nk/N}; \quad k = 0, 1, \cdots, N-1 \quad (2.28)$$

$$\text{Synthesis expression: } x(n) = \frac{1}{N} \sum_{k=0}^{N-1} X(k)e^{j2\pi nk/N}; \quad n = 0, 1, \cdots, N-1 \quad (2.29)$$

The DFT provides the spectrum of the finite length signal. The finite signal of $N$ length is equivalent to multiply the infinite length signal with the window corresponding to length of $N$ samples. In the frequency-domain, it can be considered as a convolution process between the actual spectrum of the signal and the spectrum of the window function.

It should be noted that the finite length of the signal can also be considered as rectangular windowed version of the actual signal, which is similar to the convolution between the spectrum of the signal and sinc type of the spectrum of window in frequency-domain. As a result, distortion occurs, which is known as the "spectral leakage effect." It can be reduced by the use of properly designed window [17].

In order to reduce the spectral leakage, many windows have been designed in the literature. Figure 2.4 shows three windows, namely, Gaussian, Hamming, and Kaiser, in the time- and frequency-domain [18, 19, 20].

The spectral leakage problem is shown with the help of an example of the synthetic signal $x(n) = \sin(2\pi \frac{10n}{500}) + \sin(2\pi \frac{20n}{500})$ and spectrum of its is shown in Figs. 2.5 (a) and (b). Gaussian window function is multiplied with signal $x(n)$ in order to get windowed signal, which is shown in Fig. 2.5 (c). The spectrum of the windowed signal is shown in Fig. 2.5 (d). Without use of window function, there are spectral leakages in the spectrum, which can be observed by comparing Figs. 2.5 (b) and (d). It can be noticed that the use of the Gaussian window reduces the spectral leakages. The window function is also useful for designing the filters. In order to reduce spectral leakages, the window function is applied as a weight to the signal before processing using DFT.

## 2.5  SIGNAL REPRESENTATION IN TERMS OF BESSEL FUNCTIONS

The Bessel functions are aperiodic in nature and decay with time. Such properties make Bessel functions suitable for representation of non-stationary signals [21]. The Fourier-Bessel series expansion (FBSE) based representation employs the Bessel functions as a set of basis functions. The Bessel functions are the solution of the following differential equation [22, 23]:

$$t^2 \frac{d^2 y}{dt^2} + t \frac{dy}{dt} + (t^2 - n^2)y = 0 \quad (2.30)$$

## Table 2.2
## Various Forms of Fourier Transform

| Signals | Corresponding Fourier transform |
|---|---|
| Continuous-time energy signals | Continuous-time Fourier transform (CTFT) <br><br> $: \begin{cases} x(t) : \text{Continuous-time, aperiodic} \\ X(\omega) : \text{Continuous-frequency, aperiodic} \end{cases}$ <br><br> Synthesis expression: $x(t) = \frac{1}{2\pi} \int_{-\infty}^{\infty} X(\omega) e^{j\omega t} \, d\omega$ <br><br> Analysis expression: $X(\omega) = \int_{-\infty}^{\infty} x(t) e^{-j\omega t} \, dt$ |
| Continuous-time power signals | Continuous-time Fourier series (CTFS) <br><br> $: \begin{cases} x(t) : \text{Continuous-time, periodic} \\ g_m : \text{Discrete-frequency, aperiodic} \end{cases}$ <br><br> Synthesis expression: $x(t) = \sum_{m=-\infty}^{\infty} g_m e^{jm\omega_0 t}$ <br><br> Analysis expression: $g_m = \frac{1}{T} \int_0^T x(t) e^{-jm\omega_0 t} \, dt$ <br> $T$: Period of the signal |
| Discrete-time energy signals | Discrete-time Fourier transform (DTFT) <br><br> $: \begin{cases} x(n) : \text{Discrete-time, aperiodic} \\ X(\omega) : \text{Continuous-frequency, periodic} \end{cases}$ <br><br> Synthesis expression: $x(n) = \frac{1}{2\pi} \int_0^{2\pi} X(\omega) e^{j\omega n} \, d\omega$ <br><br> Analysis expression: $X(\omega) = \sum_{n=-\infty}^{\infty} x(n) e^{-j\omega n}$ |
| Discrete-time power signals | Discrete-time Fourier series (DTFS) <br><br> $: \begin{cases} x(n) : \text{Discrete-time, periodic} \\ C_k : \text{Discrete-frequency, periodic} \end{cases}$ <br><br> Synthesis expression: $x(n) = \sum_{k=0}^{N-1} C_k e^{j2\pi nk/N}$ <br><br> Analysis expression: $C_k = \frac{1}{N} \sum_{n=0}^{N-1} x(n) e^{-j2\pi nk/N}$ <br> Frequency corresponding to $k^{\text{th}}$ point is given by $f_k = \frac{k f_s}{N}$ <br> where, $f_s$ and $N$ are sampling rate and period of the signal, respectively. |

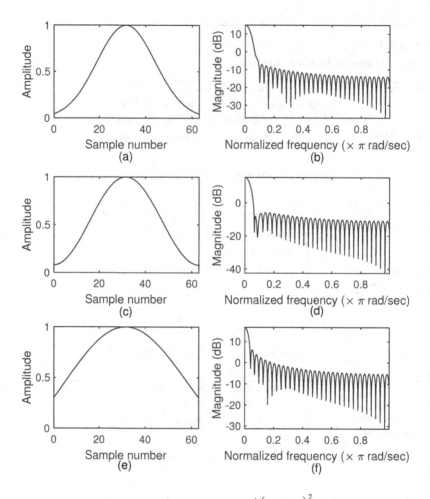

**Figure 2.4** (a) Gaussian window, $w(n) = e^{-\frac{1}{2}\left(\alpha\frac{n}{(L-1)/2}\right)^2}$, where $\alpha = 2.5$ and $L = 64$ and (b) its corresponding magnitude spectrum, (c) Hamming window, $w(n) = 0.54 - 0.46\cos\left(2\pi\frac{n}{N}\right)$, where $N = 63$ and (d) its corresponding magnitude spectrum, and (e) Kaiser window, $w(n) = \dfrac{I_0\left(\beta\sqrt{1-\left(\frac{n-N/2}{N/2}\right)^2}\right)}{I_0(\beta)}$, where $I_0(\cdot)$ represents zero-order modified Bessel function of the first kind, $\beta = 2.5$, and $N = 64$, and (f) its corresponding magnitude spectrum (for all windows $n$ varies from 0 to $N-1$, where $N$ is the window length).

The Eq. (2.30) is also termed as Bessel's equation of order $n$. The solution of Eq. (2.30) can be given in terms of the first kind of Bessel functions of order $n$ denoted by $J_n(t)$, and the second kind of Bessel functions of order $n$ denoted by $Y_n(t)$, which is expressed as,

$$y(t) = AJ_n(t) + BY_n(t) \tag{2.31}$$

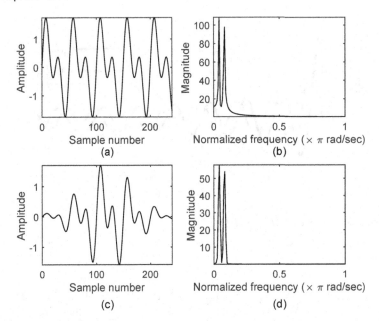

**Figure 2.5** Windowing effect on the spectrum: (a) and (b) represent a finite length signal and its magnitude spectrum, respectively, (c) and (d) are windowed finite length signal and its magnitude spectrum, respectively.

where $A$ and $B$ are the arbitrary constants. The first kind of Bessel functions of order-zero $J_0(t)$ and order-one $J_1(t)$ has been found very useful for representing non-stationary signals for many applications. The $J_0(t)$ and $J_1(t)$ Bessel functions can be mathematically expressed in the form of series expansion as follows:

$$J_0(t) = 1 - \frac{t^2}{(1!)^2 2^2} + \frac{t^4}{(2!)^2 2^4} - \frac{t^6}{(3!)^2 2^6} + \cdots$$

$$J_1(t) = \frac{t}{2} - \frac{t^3}{1!2!2^3} + \frac{t^5}{2!3!2^5} - \frac{t^7}{3!4!2^7} + \cdots$$

Figure 2.6 shows the first-kind order-zero and order-one Bessel functions.

### 2.5.1 FOURIER-BESSEL SERIES EXPANSION

The FBSE based on the first kind and zero-order Bessel functions is termed as order-zero FBSE [24, 25, 26, 27]. On the other hand, FBSE based on the first kind and one-order Bessel functions is known as order-one FBSE. Analysis and synthesis expressions for both order-zero and order-one FBSEs are shown in Table 2.3.

It should be noted that the first kind of Bessel functions of order-$n$ follow the orthogonality property in the interval $(0, a)$, which can be expressed as follows:

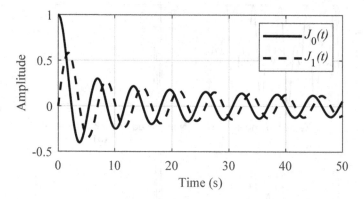

**Figure 2.6**  Plots of $J_0(t)$ and $J_1(t)$.

$$\int_0^a t J_n\left(\frac{\xi_i t}{a}\right) J_n\left(\frac{\xi_j t}{a}\right) dt = \begin{cases} \frac{a^2}{2}[J_{n+1}(\xi_i)]^2, & \text{if } \xi_i = \xi_j \\ 0, & \text{otherwise} \end{cases}$$

where $\xi_i$ and $\xi_j$ are the $i^{th}$ and $j^{th}$ roots of $J_n(x)$.

The roots of zero-order Bessel function $J_0(t)$ can be computed using the Newton-Raphson method, which is derived from Taylor series expansion [28]. For root computation, a simplified Taylor series expansion of $J_0(t)$ can be considered as follows:

$$J_0(t_{k+1}) = J_0(t_k) + (x_{t+1} - t_k)J_0'(t_k) \qquad (2.32)$$

Considering $J_0(t_{k+1}) \ll J_0(t_k)$,

$$t_{k+1} = t_k - \frac{J_0(t_k)}{J_0'(t_k)} \qquad (2.33)$$

As we know that, $J_0'(t) = -J_1(t)$. Hence,

$$t_{k+1} = t_k + \frac{J_0(t_k)}{J_1(t_k)} \qquad (2.34)$$

In order to initialize the iteration of root computation, the initial value of roots of $J_0(t)$ can be obtained using the following relation:

$$\xi_{k+1} \approx \xi_k + \pi \qquad (2.35)$$

The iteration is stopped when the value of root converges to its actual value. Similarly, the roots of the first-order Bessel function $J_1(t)$ can be computed.

The order-zero FBSE of a sinusoidal signal $x(t) = \cos(\omega t)$ can be expressed as,

$$a_k = \frac{2\lambda_k \cos(\omega a - \theta)}{J_1(\lambda_k)[(\lambda_k^2 - \omega^2 a^2)^2 + \omega^2 a^2]^{1/2}} \qquad (2.36)$$

**Table 2.3**

**Analysis and Synthesis Expressions for Order-Zero and Order-One FBSE**

| Order-Zero FBSE | Order-One FBSE |
|---|---|
| **Continuous-time signals with interval (0, a)** | |
| Synthesis expression:<br><br>$$x(t) = \sum_{k=1}^{\infty} a_k J_0\left(\frac{\xi_k t}{a}\right)$$<br>For bandlimited signals, the synthesis expression can be written as,<br>$$x(t) = \sum_{k=1}^{L} a_k J_0\left(\frac{\xi_k t}{a}\right)$$ | Synthesis expression:<br><br>$$x(t) = \sum_{k=1}^{\infty} b_k J_1\left(\frac{\xi_k t}{a}\right)$$<br>For bandlimited signals, the synthesis expression can be written as,<br>$$x(t) = \sum_{k=1}^{L} b_k J_1\left(\frac{\xi_k t}{a}\right)$$ |
| Analysis expression:<br><br>$$a_k = \frac{2}{a^2 [J_1(\xi_k)]^2} \int_0^a t x(t) J_0\left(\frac{\xi_k t}{a}\right) dt$$ | Analysis expression:<br><br>$$b_k = \frac{2}{a^2 [J_0(\xi_k)]^2} \int_0^a t x(t) J_1\left(\frac{\xi_k t}{a}\right) dt$$ |
| **Discrete-time signals with interval [0, L-1]** | |
| Synthesis expression:<br><br>$$x(n) = \sum_{k=1}^{L} A_k J_0\left(\frac{\xi_k n}{L}\right),$$<br>$$n = 0, 1, ..., L-1$$ | Synthesis expression:<br><br>$$x(n) = \sum_{k=1}^{L} B_k J_1\left(\frac{\xi_k n}{L}\right),$$<br>$$n = 0, 1, ..., L-1$$ |
| Analysis expression:<br><br>$$A_k = \frac{2}{L^2 [J_1(\xi_k)]^2} \sum_{n=0}^{L-1} n x(n) J_0\left(\frac{\xi_k n}{L}\right),$$<br>$$k = 1, ..., L$$ | Analysis expression:<br><br>$$B_k = \frac{2}{L^2 [J_0(\xi_k)]^2} \sum_{n=0}^{L-1} n x(n) J_1\left(\frac{\xi_k n}{L}\right),$$<br>$$k = 1, ..., L$$ |

where

$$\theta = \sin^{-1}\left(\frac{\omega a}{[(\lambda_k^2 - \omega^2 a^2)^2 + \omega^2 a^2]^{1/2}}\right) \qquad (2.37)$$

Further, the peak value of FBSE coefficients, $a_k$, is attained at $k^{\text{th}}$ order value, where the root, $\lambda_k \approx \omega a$ and,

$$a_{k,\text{peak}} \approx \frac{2\sin(\omega a)}{J_1(\omega a)} \qquad (2.38)$$

The value of $a_k$ decreases as we move away on either side from order $k$, and the value becomes insignificant at far away orders, $k$.

The one-to-one mapping between continuous frequency $f_k$ and order $k$ for order-zero FBSE is given by the following expressions:

$$\text{Continuous case}: \xi_k = 2\pi f_k a \qquad (2.39)$$

$$\text{Discrete case}: \xi_k = \frac{2\pi f_k L}{f_s} \tag{2.40}$$

where $f_s$ is the sampling frequency. We know that, $\xi_k = \xi_{k-1} + \pi \approx k\pi$. Hence, order $k$ for order-zero FBSE can be mapped to continuous frequency $f_k$ as follows:

$$\text{Continuous case}: f_k = \frac{k}{2a} \tag{2.41}$$

$$\text{Discrete case}: f_k = \frac{f_s k}{2L} \tag{2.42}$$

For plotting the FBSE spectrum of a signal, the FBSE coefficients can be plotted against frequency $f_k$ derived from order $k$ using Eq. (2.42). In order to get better visual representation, magnitude of the FBSE coefficients can be used instead of FBSE coefficients. For example, an AM signal $x(n) = \left[A + \mu \cos\left(\omega_0 \frac{n}{f_s}\right)\right] \cos\left(\omega_c \frac{n}{f_s}\right)$, where $A = 1$, $\mu = 0.8$, $\omega_0 = 100\pi$ rad/sec, $\omega_c = 1100\pi$ rad/sec, sampling frequency $f_s = 5000$ Hz, and $n = 1, 2, 3, \ldots, 512$ has been considered. The time-domain representation of the aforementioned signal along with order-zero FBSE spectrum is shown in Fig. 2.7.

The FBSE-based spectrum has the following advantages over Fourier-based representation:

FBSE-based representation uses the Bessel functions as a basis set, which are aperiodic and decaying with respect to time. Such characteristics of Bessel functions make them suitable for the analysis of non-stationary signal. Whereas Fourier transform uses complex exponentials (sine and cosine) functions as a basis set to represent the signal. These complex exponential basis functions are periodic and stationary in nature, which make them unsuitable for non-stationary signal analysis.

In Fourier-based analysis of signals, there is a need of window to reduce the spectral leakages in the spectrum. Whereas, in FBSE-based analysis, there is no need of window.

In Fourier-based analysis, the complex basis functions are used for analysis of real signal which makes its coefficients complex. Whereas, in FBSE-based signal analysis, real basis functions are used for analysis of real signal which provides the real coefficients.

In Fourier representation, due to the use of complex basis functions, the concept of negative frequencies arises, which does not have any physical interpretation. But, the FBSE spectrum has only positive frequency components.

The basis function set in Fourier analysis is non-convergent in nature, whereas the basis function set in FBSE analysis is convergent.

Due to the AM nature of basis function in FBSE analysis, the wide-band signals can be compactly represented by FBSE coefficients as compared to Fourier coefficients.

Due to the conjugate symmetry in Fourier coefficients for the analysis of real signals, the frequency resolution in Fourier analysis becomes half as compared to FBSE-based analysis in which there is no symmetry in the spectrum.

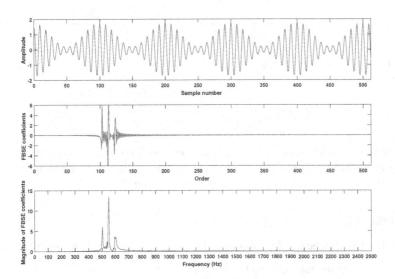

**Figure 2.7**  Plots of (a) AM signal, (b) FBSE coefficients, and (c) magnitude of FBSE coefficients.

## 2.5.2  FOURIER-BESSEL TRANSFORM

If the interval $a$ of the Bessel functions used in the FBSE is extended to the infinite interval, then discrete roots $\xi_i$ become continuous variable $\chi$ [22, 29]. Using this concept, the analysis equation for order-$n$ Fourier-Bessel transform (FBT) can be given as,

$$Y(\chi) = \int_0^\infty ty(t)J_n(\chi t)dt$$

The above expression is also termed as Hankel transform. Its inverse transform or synthesis expression can be given by,

$$y(t) = \int_0^\infty \chi Y(\chi)J_n(\chi t)d\chi$$

The considered Bessel functions are orthogonal with respect to weight factor $t$, i.e.,

$$\int_0^\infty tJ_n(\chi_i t)J_n(\chi_j t)dt = \frac{\delta(\chi_i - \chi_j)}{\chi_i} \qquad (2.43)$$

The order-zero FBT has been used for the compact representation of signal, $y(t)$, where the Bessel functions are suggested to be used as basis functions. For an example, consider a exponentially damped cosine signal $y(t) = e^{-\sigma t}\cos(\omega_0 t)$, $\sigma > 0$. The order-zero FBT of $y(t)$ can be given as [30],

$$Y(\chi) = \frac{\chi}{2}\frac{[\chi^2 + (\sigma + j\omega_0)^2]^{\frac{3}{2}} + [\chi^2 + (\sigma - j\omega_0)^2]^{\frac{3}{2}}}{[(\chi^2 + \sigma^2 - \omega_0^2)^2 + 4\sigma^2\omega_0^2]^{\frac{3}{2}}} \tag{2.44}$$

Further, it can be simplified as,

$$Y(\chi) = \frac{\chi\cos\left(\frac{3}{2}\tan^{-1}\left(\frac{-2\chi\omega_0}{\chi^2+\sigma^2-\omega_0^2}\right)\right)}{[(\chi^2+\sigma^2-\omega_0^2)^2 + 4\sigma^2\omega_0^2]^{\frac{3}{4}}} \tag{2.45}$$

The transform attains the peak value at $\chi \approx \omega_0$ and becomes zero at far away points. This mean that the variable $\chi$ can provide the information about frequency and plot of FBT ($Y(\chi)$) with respect to $\chi$ can be used as spectrum.

## PROBLEMS

Q 2.1 Check whether the following signals satisfy the Dirichlet conditions or not.

(a) cosine function: $x(t) = \cos(\omega t)$

(b) sine function: $x(t) = \sin(\omega t)$

(c) constant function: $x(t) = K$

(d) signum function: $x(t) = \text{sgn}(t) = \begin{cases} -1; & t < 0 \\ 0; & t = 0 \\ 1; & t > 0 \end{cases}$

Suggest an approach for computing the Fourier transform of these above-mentioned signals without using any property of the Fourier transform and explain graphically.

Q 2.2 For a sinusoidal signal $x(t) = \sin(\omega_0 t)$, sketch the following:

(a) Fourier spectrum of signal $x(t)$ for entire duration.

(b) Fourier spectrum of sampled version of signal $x(t)$ for entire duration.

(c) Fourier spectrum of rectangular windowed signal $x(t)$.

(d) Fourier spectrum of rectangular windowed sampled version of signal $x(t)$.

Discuss about the pattern and nature of the spectrums for above cases (a) to (d).

Q 2.3 Suppose we want to approximate signal $x(t)$ which is shown in Fig. 2.8. (a) using the given set of basis functions shown in Fig. 2.8 (b).

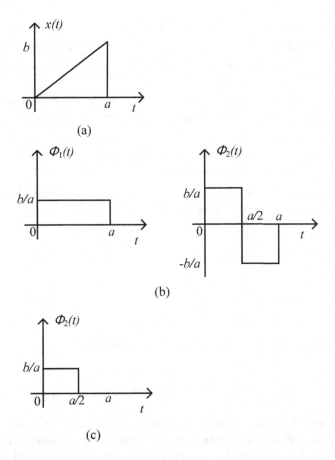

(a)

(b)

(c)

**Figure 2.8**   The set of basis functions for Q 2.3.

(a) Determine the coefficients $C_k$ so that $\hat{x}(t) = \sum_{k=1}^{K} C_k \phi_k(t)$ approximates $x(t)$ over the interval $0 < t < a$ with minimum integral square error using basis function $\phi_1(t)$ $(K = 1)$ and basis functions $\phi_1(t)$ and $\phi_2(t)$ $(K = 2)$ both, respectively.

(b) Determine the mean square error of approximated signal and original signal for both cases $(K = 1$ and $K = 2)$.

(c) Now the first basis function is fixed and $\phi_2(t)$ basis function is modified as shown in Fig. 2.8 (c). Now, repeat the above parts of the question (a) and (b) for this case.

(d) Please check orthogonality of the basis functions for both the cases and comment on variability of coefficients in each case.

Q 2.4  Determine the Fourier transform of the following signals:

(a) Gabor filter: $x(t) = e^{\frac{-t^2}{a}} \cos(\omega_0 t)$

(b) Daughter wavelet: $\psi_{a,b}(t) = \frac{1}{\sqrt{a}} \psi\left(\frac{t-b}{a}\right)$ (Given that: Fourier transform of mother wavelet $\psi(t)$ is $\Psi(\omega)$), where $a$ and $b$ denote scale and shifting parameters, respectively.

(c) Damped sinusoidal signal: $x(t) = e^{-\sigma t} \cos(\omega_0 t)$, where $\sigma$ and $\omega_0$ denote the damping factor and the center frequency, respectively.

(d) AM signal: $x(t) = A[1 + \mu \sin(\omega_m t)] \cos(\omega_c t)$, where $A$ is carrier amplitude, $\mu$ denotes modulation index, $\omega_m$ is message signal frequency, and $\omega_c$ is carrier signal frequency.

Q 2.5  Suppose a signal is represented by $x(t) = A\cos(\omega_0 t)$, determine the Fourier transform of the following signals:

(a) $x(t)$

(b) $x(t-b)$

(c) $x(p(t-b))$

(d) $x(pt-1)$

where $A$, $\omega_0$, $b$, and $p$ are constant parameters.

Q 2.6  Consider a pulse signal which is represented by,

$$x(t) = \begin{cases} 1, & \frac{-T}{2} \leq t \leq \frac{T}{2} \\ 0, & \text{otherwise} \end{cases}$$

where $T$ denotes the width, now this pulse signal is compressed by a factor of 4, then what will be the change in the bandwidth (bandwidth is defined by spread of the first lobe between first zero crossing in the positive frequency range and first zero crossing in the negative frequency range). Explain this effect with the sketch of the spectrum for both cases.

Q 2.7  Consider three LTI systems with impulse responses $h_1(t)$, $h_2(t)$, and $h_3(t)$ are cascaded together. Find out the output signal $y(t)$ of the resultant system for input signal $x(t)$.
where,

$$x(t) = \sum_{k=-\infty}^{\infty} \delta(t-kT)$$

$$h_i(t) = \frac{\sin(\beta_i t)}{\pi t}, \quad \text{with } \beta_1 = B,\ \beta_2 = \frac{B}{2},\ \beta_3 = \frac{B}{4},\ \text{and } i = 1,2,3.$$

Q 2.8  Hilbert transform of $x(t)$ is defined by $x_H(t) = x(t) * \frac{1}{\pi t}$. With the help of Fourier transform, determine the Hilbert transform of the following signal:

$$y(t) = \sum_{k=1}^{K} A_k e^{j\omega_k t}$$

Q 2.9  A signal bandlimited to $f_m$ Hz whose spectrum is shown in Fig. 2.9 sampled as per sampling theorem. The sampling rate is considered $f_s = 8f_m$ Hz. Sketch the spectrum of the sampled signal. Suppose an ideal low-pass filter is applied to the sampled signal, whose cutoff frequency varies as follows:

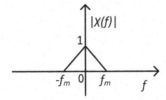

**Figure 2.9**   The spectrum of signal for Q 2.9.

(a)  $f_c = \frac{f_s}{4}$

(b)  $f_c = \frac{f_s}{8}$

(c)  $f_c = \frac{f_s}{16}$

Explain which of the above-mentioned cases the signal can be reconstructed without distortion.

Q 2.10  A signal $g(t) = 3\cos(200\pi t) + 4\cos(300\pi t)$ is sampled at sampling rate of 300 Hz, and this sampled signal is passed through an ideal low-pass filter, whose cutoff frequency is 110 Hz. Determine the output signal and sketch Fourier and FBSE spectrums of the output signal.

Q 2.11  Consider a signal $x(t) = \cos(10\pi t) + \cos(12\pi t)$ and sample it with sampling rate of 100 Hz. Determine the Fourier spectrum of signal for signal lengths of 50, 100, and 200 samples. Explain the resolution of spectrum with respect to these considered signal lengths. Compare the Fourier spectrum with the FBSE spectrum for the same signal of different lengths.

Q 2.12  A signal $x(t) = \cos(300\pi t) + \cos(600\pi t)$ is sampled at sampling rate of 2000 Hz. Determine the Fourier spectrum corresponding to the signal length of 512 samples and 1024 samples (out of these 1024 samples, 512 samples are taken from signal and remaining 512 samples are obtained by zero padding). Show the effect of the zero-padding process in the obtained frequency components by Fourier representation.

# 3 Basics of Time-Frequency Analysis

*"It is not the strongest of the species that survives, nor the most intelligent. It is the one most adaptable to change."* –Charles Darwin

## 3.1 TIME-DOMAIN REPRESENTATION

In time-domain representation of the signal, generally amplitude of the signals is plotted against time parameter. For example, speech signal shown in Fig. 3.1, which represents variation of amplitude of speech signal with respect to time. The sampling frequency of the signal is 32000 Hz. This signal is taken from the CMU ARCTIC database [31].

In signal processing, we are interested in representing the unknown or complicated signal in terms of simple sinusoidal signals like this set $\{A_k\cos(\omega_k t)\}$, which represents sinusoidal signal of amplitude $A_k$ and frequency $\omega_k$. The harmonicaly related sinusoidal functions are expressed as, $\{A_k\cos(k\omega_0 t)\}$. Such kind of sinusoidal functions can be represented by only two parameters $\{A_k, \omega_k\}$ or in case of harmonically related sinusoidal functions $\{A_k, k\omega_0\}$, where $\omega_0$ is the fundamental frequency of sinusoidal signal. An example is shown in Fig. 3.2 where three harmonics of 50 Hz sinusoidal signal are sampled at 8000 Hz sampling rate.

Generalization of the signal representation in terms of sinusoidal functions leads to amplitude and frequency modulated (AFM) signal model, where each sinusoidal component can have time-varying amplitude and frequency in the representation. Single or mono-component AFM component is defined as follows:

$$x(t) = a(t)\cos[\phi(t)] \tag{3.1}$$

Where $a(t)$ and $\phi(t)$ are functions of time and this signal can be represented by pair of these two functions $\{a(t), \phi(t)\}$.

An example of mono-component AFM signal is shown in Fig. 3.3, where $a(t) = 0.2 + \sin(0.25\pi t)$, $\phi(t) = 10\pi t + 15t^2$, and AFM signal $x(t) = [0.2 + \sin(0.25\pi t)] \times \cos(10\pi t + 15t^2)$. The sampling frequency of the signals is considered as 10000 Hz.

The multi-component AFM signal model can be expressed as follows:

$$x(t) = \sum_{i=1}^{M} a_i(t)\cos[\phi_i(t)] = \sum_{i=1}^{M} x_i(t) \tag{3.2}$$

This AFM model has been useful in communication system for transformation of the information. It is also used to represent the wide-band non-stationary signal in terms of narrow-band non-stationary signals like speech. The various parametric forms of the AFM model are also developed in the literature.

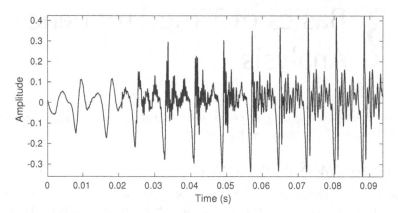

**Figure 3.1**   Speech signal.

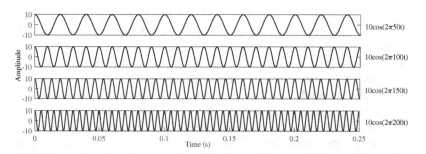

**Figure 3.2**   Sinusoidal signal of 50 Hz and its corresponding three harmonics.

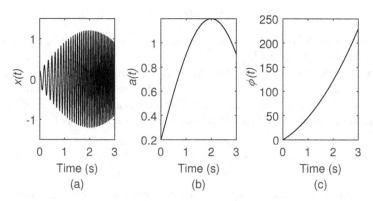

**Figure 3.3**   (a) Mono-component AFM signal $a(t)\cos[\phi(t)]$, (b) amplitude envelope (AE) $a(t)$, and (c) time-varying phase function $\phi(t)$.

### 3.1.1 THE NON-PARAMETRIC AFM SIGNAL MODEL

The non-parametric AFM signal model in general form can be represented by,

$$x(t) = e(t)\cos\left[\omega_c t + \omega_m \int_0^t p(\tau)d\tau + \theta\right] \tag{3.3}$$

where angle $\theta(t) = \omega_c t + \omega_m \int_0^t p(\tau)d\tau + \theta$. The instantaneous frequency (IF), which is derivative of the phase can be expressed as,

$$\omega(t) = \frac{d\theta(t)}{dt} = \omega_c + \omega_m p(t) \tag{3.4}$$

Here, $|p(t)| \leq 1$ and $\omega_m$ is the maximum frequency deviation with respect to $\omega_c$ and $\theta = \theta(0)$.

The speech formant is modeled using exponentially damped AFM signal model. The exponentially damped AFM signal model can be expressed as,

$$x(t) = e^{-\sigma t}e(t)\cos\left[\omega_c t + \omega_m \int_0^t q(\tau)d\tau + \theta\right] \tag{3.5}$$

The complete speech signal can be represented as a sum of such AFM signals and can be expressed as,

$$s(t) = \sum_{k=1}^{K} e_k(t)\cos[\theta_k(t)] \tag{3.6}$$

Here $k$ denotes the $k^{\text{th}}$ resonance of speech signal and $K$ is the number of total speech formants. Now, estimation of $\{e_k(t), \theta_k(t) \text{ or } \frac{d\theta_k(t)}{dt} = \omega_k(t)\}$ is a challenging task in this modeling. As a solution, Gabor band-pass filter has been suggested to use in order to separate these speech formants or resonances. The impulse response of Gabor band-pass filter is given by,

$$g(t) = e^{-\alpha^2 t^2}\cos(\omega_0 t) \tag{3.7}$$

In frequency-domain, it can be expressed as,

$$G(\omega) = \frac{\sqrt{\pi}}{2\alpha}\left(e^{-\frac{(\omega-\omega_0)^2}{4\alpha^2}} + e^{-\frac{(\omega+\omega_0)^2}{4\alpha^2}}\right) \tag{3.8}$$

The Gabor filter has some advantages over other filtering techniques. It is optimally compact in time- and frequency-domain and it has no side lobes in the spectrum.

This type of formants separation suffers from some problems such as selections of center frequency and bandwidth of the filter and introduction of the modulation in the separated speech formants. The discrete energy separation algorithm (DESA) has been used to estimate amplitude envelope (AE) and IF of these separated formants. The mathematical expressions for DESA methods for discrete-time AFM signal defined in Eq. (3.9) are given as [32],

$$s(n) = e(n)\cos\left[\omega_c n + \omega_m \int_0^n p(m)dm + \theta\right] \tag{3.9}$$

where IF: $\omega(n) = \frac{d\theta(n)}{dn} = \omega_c + \omega_m p(n)$.

DESA-1a: Where '1' indicates the approximation of derivatives with a single sample difference and 'a' denotes to the usage of asymmetric difference. $y(n)$ is defined as the approximation of first derivative of signal $s(n)$ and can be expressed as,

$$y(n) = s(n) - s(n-1) \tag{3.10}$$

The estimation of IF and AE using DESA-1a is as follows:

$$\arccos\left(1 - \frac{\Psi[y(n)]}{2\Psi[s(n)]}\right) \approx \omega(n) \tag{3.11}$$

$$\sqrt{\frac{\Psi[s(n)]}{1 - \left(1 - \frac{\Psi[y(n)]}{2\Psi[s(n)]}\right)^2}} \approx |e(n)| \tag{3.12}$$

where $\Psi[\cdot]$ is a Kaiser nonlinear energy tracker operator which can be defined as,

$$\Psi[y(n)] \triangleq y^2(n) - y(n-1)y(n+1) \tag{3.13}$$

DESA-1: DESA-1a is further improved by using symmetric difference. The $z(n)$ is the approximation of first forward derivative of the signal $s(n)$ and expressed as,

$$z(n) = s(n+1) - s(n) = y(n+1) \tag{3.14}$$

The estimation of IF and AE functions using DESA-1 approach can be carried out as follows:

$$\arccos\left(1 - \frac{\Psi[y(n)] + \Psi[y(n+1)]}{4\Psi[s(n)]}\right) \approx \omega(n) \tag{3.15}$$

$$\sqrt{\frac{\Psi[s(n)]}{1 - \left(1 - \frac{\Psi[y(n)] + \Psi[y(n+1)]}{4\Psi[s(n)]}\right)^2}} \approx |e(n)| \tag{3.16}$$

It should be noted that the frequency estimation works properly for the range, $0 < \omega(n) < \pi$.

DESA-2: The $y(n)$ is approximation of first symmetric derivative of signal $s(n)$ and expressed as,

$$y(n) = [s(n+1) - s(n-1)]/2 \tag{3.17}$$

The estimation of IF and AE functions using DESA-2 can be expressed as follows:

$$\frac{1}{2}\arccos\left(1 - \frac{\Psi[2y(n)]}{2\Psi[s(n)]}\right) \approx \omega(n) \tag{3.18}$$

$$\frac{2\Psi[s(n)]}{\sqrt{\Psi[2y(n)]}} \approx |e(n)| \tag{3.19}$$

In this case, the frequency estimation part works in the range, $0 < \omega(n) < \pi/2$.

Both DESA-1 and DESA-2 perform better than DESA-1a. DESA-1 performs slightly better than DESA-2, whereas DESA-2 is fastest of all three and also its mathematical analysis is the simplest. For example, an AFM signal $x(n)$ represented in Eq. (3.20) has been considered and its AE and IF functions are computed using DESA-1 algorithm. The AE and IF functions of signal $x(n)$, computed using DESA-1 algorithm, is shown in Fig. 3.4.

$$x(n) = \left[1 + 0.4\cos\left(30\pi\frac{n}{f_s}\right)\right]\cos\left[2\pi\left(550 + 1250\frac{n}{f_s}\right)\frac{n}{f_s}\right] \tag{3.20}$$

where $f_s$ is sampling frequency of the signal which is equal to 5000 Hz.

In order to overcome the limitations of the Gabor filtering and DESA-based method, the FBSE-based DESA (FB-DESA) has been proposed, which is order-zero FBSE-based DESA-1 algorithm [33]. The FB-DESA technique requires manual indentation of the FBSE coefficients for separation of speech formants. This method does not require prior information about frequency bands of the mono-component signals or formants of the speech signals.

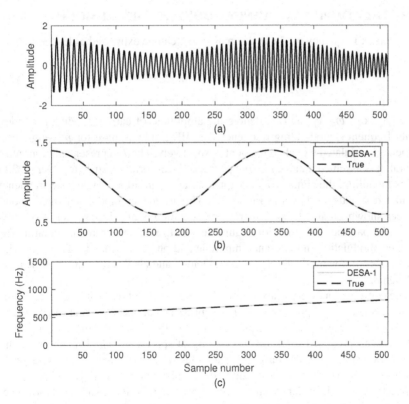

**Figure 3.4** (a) Time-domain signal $x(n)$, (b) AE and (c) IF functions, computed using DESA-1.

### 3.1.2  THE COMPLEX AMPLITUDE MODULATED SIGNAL MODEL

The complex amplitude modulated signal model is given as [34, 35],

$$x(n) = \sum_{m=1}^{M} A_m[1 + \mu_m e^{jv_m nT}] e^{j\omega_m nT} e^{j\phi_m} \qquad (3.21)$$

The amplitude modulated signal model has been found suitable for parametric representation of voiced speech signals efficiently. Where $A_m$ is the carrier amplitude, $\mu_m$ is the modulation index, $\omega_m$ is the carrier frequency, $v_m$ is the modulating frequency, and $\phi_m$ is the independent and identically distributed (IID) random phase uniformly distributed over $[0,2\pi)$ for $m^{\text{th}}$ single-tone amplitude modulated signal. The $T$ denotes the sampling interval. The signal $x(n)$ is represented as a sum of $M$ single-tone amplitude modulated signals as shown in Eq. (3.21). The accumulated autocorrelation function of the modeled signal is used in order to estimate the parameters of the model.

### 3.1.3  THE COMPLEX FREQUENCY MODULATED SIGNAL MODEL

The complex frequency modulated signal model can be expressed as [34, 36],

$$x(n) = \sum_{p=1}^{P} A_p e^{j[\omega_p nT + \phi_p + \alpha_p \sin(v_p nT)]} \qquad (3.22)$$

Where $A_p$, $\alpha_p$, $\omega_p$, $v_p$, and $\phi_p$ denote the amplitude of carrier, modulation index, carrier frequency, modulating frequency, and IID random phase for $p^{\text{th}}$ single-tone frequency modulated signal component, respectively. The $T$ represents the sampling interval. It is assumed that the random phase, $\phi_p$ is uniformly distributed over $[0, 2\pi)$. In this modeling, the signal $x(n)$ is expressed as a sum of $M$ single-tone frequency modulated signals as shown in Eq. (3.22). This frequency modulated signal model has been shown suitable for parametric representation of unvoiced speech phonemes. The model parameters estimation requires estimation of carrier and modulating frequencies, modulation indices, and amplitude and phase parameters. The estimation approaches employ product function computation and DFT computation of the modeled signal.

The combination of complex amplitude modulated and frequency modulated signal models has been applied in order to represent real speech signal in parametric manner. These complex amplitude modulated and frequency modulated signal models have been proposed for online processing of speech signals. It is observed that some voiced speech signals have characteristics similar to unvoiced speech signals. Therefore, in some cases, it is difficult to distinguish voiced and unvoiced phonemes in speech signals, in order to apply complex amplitude modulated and frequency modulated signal models for parametric representation. Based on this motivation, an AFM signal model-based parametric representation has been proposed that can represent both voiced and unvoiced speech signals.

## 3.1.4 THE PARAMETRIC AFM SIGNAL MODEL

The real single-component discrete-time AFM signal model can be given by [37],

$$x(n) = A\left[1 + \sum_{m=1}^{M} \mu_m \cos(\omega_m n + \phi_m)\right] \cos\left[\omega n + \sum_{p=1}^{P} \beta_p \sin(\omega_p n + \phi_p)\right] \quad (3.23)$$

Where $A$ and $\omega$ represent amplitude and carrier frequency of the signal, $\omega_m$ and $\omega_p$ denote modulating angular frequencies, $\phi_m$ and $\phi_p$ represent the modulating phases of modulated signal, $\mu_m$ and $\beta_p$ denote modulation indices of amplitude modulated and frequency modulated parts, respectively. The subscript parameters $p$ and $m$ represent the $p^{th}$ and $m^{th}$ tone frequency modulating and amplitude modulating parameters, respectively. $M$ and $P$ denote the total number of tones in amplitude modulated and frequency modulated parts, respectively.

In order to apply AFM signal model in speech, phoneme separation is carried out using phoneme boundary detection. The speech phoneme is modeled by multicomponent AFM signal model as,

$$x(n) = \sum_{k=1}^{K} x_k(n), \quad n = 0, 1, 2, \ldots, N-1 \quad (3.24)$$

Where, $x_k(n)$ is the $k^{th}$ AFM signal component as defined in Eq. (3.23). The order-zero FBSE method is used to separate all the components of this model for signal analysis. The DESA-1 technique has been used to obtain AE and IF functions. Estimation of the amplitude modulated signal parameters and the amplitude of the signal is carried out by analyzing the AE function, whereas the estimation of frequency modulated signal parameters and carrier frequency of the signal is performed from the analysis of IF function.

Another parametric AFM model has been proposed for speech signal analysis which is defined as follows [38]:

$$\begin{aligned} x(n) = &A\cos\left[\omega_c n + k_f \sin(\omega_f n) + \phi\right] \\ &+ \frac{pAk_a}{2}\cos\left[\omega_c n + \omega_a n + k_f \sin(\omega_f n) + \phi + \phi_a\right] \\ &+ \frac{qAk_a}{2}\cos\left[\omega_c n - \omega_a n + k_f \sin(\omega_f n) - \phi - \phi_a - \phi_b\right] \end{aligned} \quad (3.25)$$

Where, $A$ is the amplitude of the carrier signal, $\omega_c$ is the carrier frequency, $\omega_f$ and $\omega_a$ denote the modulating frequencies of the frequency modulated and amplitude modulated parts, respectively, $k_f$ and $k_a$ are the modulation indices of the frequency modulated and amplitude modulated parts, respectively, $\phi$ is the phase of the carrier assumed to be uniformly distributed random variable over $[0, 2\pi)$, $\phi_a$ is the phase of the amplitude modulated part, $\phi_b$ is the additional phase of the lower side band of the amplitude modulated part, and $p$ and $q$ are the scaling factors for the upper and lower side bands of the amplitude modulated part, respectively.

Multi-component speech signal is separated into mono-component signals using the order-zero FBSE technique. The lowest frequency or first mono-component signal can be modeled using Eq. (3.25). For the higher frequency mono-component signals, multiple sets of parameters are required, where each set of parameters is for different segments of the signal components. Based on this approach, speech signal can be modeled for all phonemes.

## 3.2 TIME-DOMAIN LOCALIZATION

The energy of the signal represents the amount of work which is required to reproduce the signal. This is the basic concept in signal processing and strength of the signal is generally measured in terms of energy. Energy has been defined in many ways in electromagnetic theory, acoustics, circuits, etc. In all these definitions, the square of the different measures have been considered as an energy. In the similar way, the energy of the signal is defined by its squaring operation [39]. Therefore, total energy of the signal $s(t)$ is given by,

$$E = \int |s(t)|^2 dt \tag{3.26}$$

In order to measure the localization of the signal in time-domain or signal center with its spread with respect to that time can be determined by two parameters namely, average (mean) time $m_t$ or standard deviation $\sigma_t$ (shown in Fig. 3.5 (a)). The mathematical expressions for mean time and standard deviation are given by,

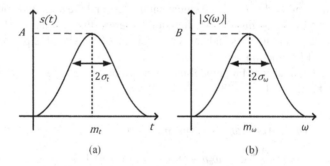

(a)                                     (b)

**Figure 3.5**   (a) Localization of signal in time-domain ($A$ is the value of $s(t)$ at $m_t$) and (b) frequency-domain localization of the signal ($B$ is the value of $|S(\omega)|$ at $m_\omega$).

$$m_t = \frac{1}{E} \int_{-\infty}^{\infty} t |s(t)|^2 dt \tag{3.27}$$

$$\sigma_t = \sqrt{\frac{1}{E} \int_{-\infty}^{\infty} (t - m_t)^2 |s(t)|^2 dt}$$ 

$$= \sqrt{m_{t2} - (m_t)^2} \tag{3.28}$$

These two parameters $m_t$ and $\sigma_t$ are useful for localization of the signal in time-domain. The parameter $\sigma_t$ determines the spread of the signal which is localized around its $m_t$. The quantity $2\sigma_t$ is termed as duration of the signal. It should be noted that, a few authors have considered $\sigma_t$ as duration in literature. For an example, a rectangular pulse signal (shown in Fig. 3.6) has been considered for computing $m_t$ and $\sigma_t$. The mathematical expression for this rectangular pulse is given as,

$$s(t) = \begin{cases} B, & -t_o \leq t \leq t_o \\ 0, & \text{otherwise} \end{cases} \tag{3.29}$$

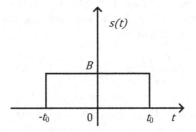

**Figure 3.6** Rectangular pulse.

Computation of $m_t$ and $\sigma_t$ is carried out as follows:

$$E = \int_{-t_0}^{t_0} B^2 dt = 2t_0 B^2$$

$$m_t = \frac{1}{2t_0 B^2} \int_{-t_0}^{t_0} tB^2 dt = B^2 \frac{t^2}{2} \Big|_{-t_0}^{t_0} = 0$$

$$\sigma_t^2 = \frac{1}{2t_0 B^2} \int_{-t_0}^{t_0} (t-0)^2 |B|^2 dt$$

$$\sigma_t^2 = \frac{1}{2t_0} \frac{t^3}{3} \Big|_{-t_0}^{t_0} = \frac{t_0^2}{3}$$

Which provides, $\sigma_t = \dfrac{t_0}{\sqrt{3}}$

## 3.3  FREQUENCY-DOMAIN LOCALIZATION

Time-domain gives the description about the signal with respect to time. In many cases, the information in frequency-domain is also required. Similar to localization parameters in time-domain, the center frequency or mean frequency and standard deviation or bandwidth of the signal are used in order to localize it in frequency-domain. These parameters provide information about the center of the concentration of frequency-domain and spread of this concentration in frequency-domain, as shown in Fig. 3.5 (b).

The Fourier transform can be used to obtain signal in frequency-domain from the time-domain. The signal in the time-domain and in frequency-domain can be represented as follows:

$$\text{Time-domain}: s(t) = \frac{1}{2\pi} \int_{-\infty}^{\infty} S(\omega) e^{j\omega t} d\omega \tag{3.30}$$

$$\text{Frequency-domain}: S(\omega) = \int_{-\infty}^{\infty} s(t) e^{-j\omega t} dt \tag{3.31}$$

It should be noted that $S(\omega)$ is a complex quantity. Hence, the magnitude of $S(\omega)$, $|S(\omega)|$, is used to represent it. Similar to time-domain, energy is defined with the help of squared magnitude of $S(\omega)$ as,

$$E = \frac{1}{2\pi} \int_{-\infty}^{\infty} |S(\omega)|^2 d\omega = \int_{-\infty}^{\infty} |s(t)|^2 dt \tag{3.32}$$

Based on the Parseval's theorem, this energy is also equal to the energy defined in time-domain as shown in Eq. (3.32). The mathematical expressions for center frequency (mean frequency) $m_\omega$ and standard deviation $\sigma_\omega$ are given as follows:

$$m_\omega = \frac{1}{E} \int_{-\infty}^{\infty} \omega |S(\omega)|^2 d\omega \tag{3.33}$$

$$\sigma_\omega = \sqrt{\frac{1}{E} \int_{-\infty}^{\infty} (\omega - m_\omega)^2 |S(\omega)|^2 d\omega} \tag{3.34}$$

$$= \sqrt{m_{\omega^2} - (m_\omega)^2}$$

Similar to the duration of the signal, the bandwidth of the signal is defined as $2\sigma_\omega$. As mentioned previously for duration of the signal, a few literatures have considered $\sigma_\omega$ as bandwidth of the signal.

## 3.4   HEISENBERG BOX REPRESENTATION

The four parameters computed from signal and its Fourier transform help to describe the Heisenberg box in time-frequency domain. The two parameters, namely, time center and frequency center represent the center of Heisenberg box. The other two parameters, namely, standard deviations in time-domain and frequency-domain represent width and height of this box, respectively [40]. The Heisenberg box is also termed as time-frequency tile. It should be noted that the width and height of this tile are governed by the Heisenberg principle. The area of the box determines the resolution for a given time-frequency representation (TFR). The minimum area is the indication of the better resolution in time-frequency domain. This box is represented by $[(m_t, m_\omega); (2\sigma_t, 2\sigma_\omega)]$. Figure 3.7 shows the sketch of Heisenberg box or time-frequency tile on time frequency plane. Table 3.1 shows the effect of shift, modulation, and scaling of the signal on Heisenberg box parameters.

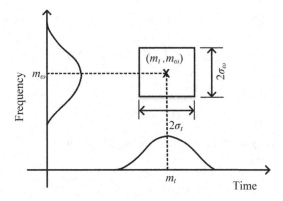

**Figure 3.7** Plot of the Heisenberg box in time-frequency plane.

**Table 3.1**

**Effect of Shift, Modulation, and Scaling Properties of the Signal on Heisenberg Box**

| Signal | Heisenberg Box |
|--------|----------------|
| $s(t)$ | $[(m_t, m_\omega); (2\sigma_t, 2\sigma_\omega)]$ |
| $s(t - t_0)$ | $[(m_t + t_0, m_\omega); (2\sigma_t, 2\sigma_\omega)]$ |
| $s(t)e^{j\omega_0 t}$ | $[(m_t, m_\omega + \omega_0); (2\sigma_t, 2\sigma_\omega)]$ |
| $\frac{1}{\sqrt{a}} s(\frac{t}{a})$ | $[(am_t, \frac{m_\omega}{a}); (2a\sigma_t, \frac{2\sigma_\omega}{a})]$ |
| $s(t - t_0)e^{j\omega_0 t}$ | $[(m_t + t_0, m_\omega + \omega_0); (2\sigma_t, 2\sigma_\omega)]$ |
| $\frac{1}{\sqrt{a}} s(\frac{t-b}{a})$ | $[(am_t + b, \frac{m_\omega}{a}); (2a\sigma_t, \frac{2\sigma_\omega}{a})]$ |

## 3.5  AM AND FM BANDWIDTHS

The computation of the mean frequency and bandwidth requires computation of spectrum of the signal. But these parameters can also be computed from the time-domain representation of the signal. Consider a complex signal $s(t)$, which is expressed as follows:

$$s(t) = A(t)e^{j\phi(t)} \tag{3.35}$$

Total energy of the signal is given by,

$$E = \int_{-\infty}^{\infty} |s(t)|^2 dt = \int_{-\infty}^{\infty} A^2(t) dt \tag{3.36}$$

Mean time ($m_t$) is computed as,

$$m_t = \int_{-\infty}^{\infty} t|s(t)|^2 dt = \int_{-\infty}^{\infty} tA^2(t) dt \tag{3.37}$$

The standard deviation ($\sigma_t$) can be computed by following expression:

$$\sigma_t = \sqrt{\int_{-\infty}^{\infty} (t - m_t)^2 A^2(t)dt} \tag{3.38}$$

The IF is computed as,

$$\omega_i(t) = \phi'(t) = \frac{d\phi(t)}{dt} \tag{3.39}$$

The mean frequency or average frequency ($m_\omega$) is given by,

$$m_\omega = \frac{\int \omega_i(t) A^2(t)\, dt}{\int A^2(t)\, dt} = \frac{1}{E} \int \phi'(t) A^2(t)dt$$

The mean frequency $m_\omega$ can also be expressed as,

$$m_\omega = \frac{1}{E} \int \omega |S(\omega)|^2 d\omega$$

Alternatively, the $m_\omega$ is also computed using $s(t)$ and its time-derivative as,

$$= \frac{1}{E} \int s^*(t) \frac{1}{j} \frac{d}{dt} s(t)dt \tag{3.40}$$

In the similar manner $m_{\omega^2}$ is also computed as follows:

$$m_{\omega^2} = \frac{1}{E} \int \omega^2 |S(\omega)|^2 d\omega = \frac{1}{E} \int s^*(t) \left(\frac{1}{j}\frac{d}{dt}\right)^2 s(t)dt$$

$$= -\frac{1}{E} \int s^*(t) \frac{d^2}{dt^2} s(t)dt$$

$$= \frac{1}{E} \int \left|\frac{ds(t)}{dt}\right|^2 dt \tag{3.41}$$

The bandwidth can be computed with the help of standard deviation $\sigma_\omega$. Let us consider computation of $\sigma_\omega^2$ with the help of $m_\omega$ as follows:

$$\sigma_\omega^2 = \frac{1}{E} \int (\omega - m_\omega)^2 |S(\omega)|^2 d\omega$$

which can be written as,

$$\sigma_\omega^2 = \frac{1}{E} \int s^*(t) \left(\frac{1}{j}\frac{d}{dt} - m_\omega\right)^2 s(t)dt$$

and can be re-written as,

$$\sigma_\omega^2 = \frac{1}{E} \int \left|\left(\frac{1}{j}\frac{d}{dt} - m_\omega\right)s(t)\right|^2 dt \tag{3.42}$$

It should be noted that,

$$m_\omega = \frac{1}{E} \int \omega |S(\omega)|^2 d\omega = \frac{1}{E} \int s^*(t) \frac{1}{j} \frac{d}{dt} s(t) dt$$

By putting $s(t) = A(t) e^{j\phi(t)}$, we have,

$$m_\omega = \frac{1}{E} \int s^*(t) \frac{1}{j} \frac{d}{dt} A(t) e^{j\phi(t)} dt$$

$$= \frac{1}{E} \int s^*(t) \frac{1}{j} \left[ A(t) j\phi'(t) e^{j\phi(t)} + e^{j\phi(t)} A'(t) \right] dt$$

$$= \frac{1}{E} \int s^*(t) s(t) \frac{1}{j} \left[ j\phi'(t) + \frac{A'(t)}{A(t)} \right] dt$$

$$= \frac{1}{E} \int A^2(t) \left[ \phi'(t) - j \frac{A'(t)}{A(t)} \right] dt$$

$$= \frac{1}{E} \int \left[ \phi'(t) + \frac{1}{j} \frac{A'(t)}{A(t)} \right] A^2(t) dt \tag{3.43}$$

Therefore,

$$\sigma_\omega^2 = \frac{1}{E} \int \left| \frac{1}{j} \frac{A'(t)}{A(t)} + \phi'(t) - m_\omega \right|^2 A^2(t) dt$$

Bandwidth is a real quantity. Therefore, imaginary part can be ignored and we have,

$$\sigma_\omega^2 = \frac{1}{E} \int \left[ \frac{A'(t)}{A(t)} \right]^2 A^2(t) dt + \frac{1}{E} \int \left[ \phi'(t) - m_\omega \right]^2 A^2(t) dt$$

The above equation can be written in following form:

$$4\sigma_\omega^2 = 4 \frac{1}{E} \int \left( \frac{A'(t)}{A(t)} \right)^2 A^2(t) dt + 4 \frac{1}{E} \int \left[ \phi'(t) - m_\omega \right]^2 A^2(t) dt \tag{3.44}$$

This expression can be written in terms of bandwidth as,

$$B_T^2 = B_{AM}^2 + B_{FM}^2$$

Here, $B_T$, $B_{AM}$, and $B_{FM}$ denote total bandwidth, bandwidth due to the amplitude modulation (AM) part, and bandwidth due to the frequency modulation (FM) part, respectively.

The total bandwidth $B_T$ can be computed as,

$$B_T = \sqrt{B_{AM}^2 + B_{FM}^2} \tag{3.45}$$

where,

$$B_{AM} = 2 \sqrt{\frac{1}{E} \int (A'(t))^2 dt} \tag{3.46}$$

and

$$B_{FM} = 2\sqrt{\frac{1}{E}\int (\phi'(t) - m_\omega)^2 A^2(t)dt} \tag{3.47}$$

The above expressions are termed as bandwidth equations. Total bandwidth of the signal depends on the AE and IF functions. Therefore, bandwidth of the signal can be varied by introducing AM, FM, or both AM and FM in the signal.

## 3.6   SPECTRUM AM AND PM DURATIONS

In similar way, the duration and mean time can also be expressed in terms of frequency-domain representation. The spectrum $S(\omega)$ of the signal $s(t)$ is given by,

$$S(\omega) = B(\omega)e^{j\phi(\omega)} \tag{3.48}$$

Energy of the signal is defined in frequency-domain by,

$$E = \int |S(\omega)|^2 d\omega \tag{3.49}$$

Mean time $m_t$ can be computed as,

$$m_t = -\frac{1}{E}\int \phi'(\omega)|S(\omega)|^2 d\omega \tag{3.50}$$

Duration of the signal can be computed based on the following expression:

$$4\sigma_t^2 = \frac{4}{E}\int [B'(\omega)]^2 d\omega + \frac{4}{E}\int [\phi'(\omega) + m_t]^2 B^2(\omega)d\omega \tag{3.51}$$

Similar to the bandwidth equation, the duration equation can be written as,

$$T_T^2 = T_{SAM}^2 + T_{SPM}^2 \tag{3.52}$$

Here, $T_T$, $T_{SAM}$, and $T_{SPM}$ denote total time duration, duration due to the spectrum AM (SAM), and duration due to the spectrum phase modulation (PM) (SPM), respectively.

Here, the average time is computed by performing averaging quantity $-\phi'(\omega)$ over all the frequency components. The same concept has been used to obtain average frequency ($m_\omega$) based on IF ($\omega_i$).

The average time for a particular frequency is called group delay denoted by $\tau_g(\omega)$, which can be expressed as,

$$\tau_g(\omega) = -\phi'(\omega) \tag{3.53}$$

It should be noted that AM and FM parts of the signal in time-domain contribute towards bandwidth. Similarly, amplitude and phase variations of the spectrum contribute towards duration of the signal. The contributions in the duration can be computed as follows:

$$T_{SAM} = 2\sqrt{\frac{1}{E}\int [B'(\omega)]^2 d\omega} \tag{3.54}$$

$$T_{SPM} = 2\sqrt{\frac{1}{E}\int [\phi'(\omega) + m_t]^2 B^2(\omega)d\omega} \qquad (3.55)$$

The total time duration is computed as,

$$T_T = \sqrt{T_{SAM}^2 + T_{SPM}^2}$$

## 3.7　UNCERTAINTY PRINCIPLE

According to uncertainty principle, for a function $x(t) \in L^2(\mathbb{R})$, the product of square of standard deviation in time-domain $(\sigma_t)$ and standard deviation in frequency-domain $(\sigma_\omega)$ is related as,

$$\sigma_t^2 \sigma_\omega^2 \geq \frac{1}{4} \qquad (3.56)$$

In this relation, the lower bound or equality is obtained by Gaussian function which can be mathematically expressed as,

$$x(t) = \alpha e^{-\beta t^2}, \quad \beta > 0 \qquad (3.57)$$

This principle can be proved using Cauchy-Schwarz inequality as follows:
Assume that $x(t)$ is a real signal and centered around $t = 0$ and it has unit energy, these assumptions are made for simplicity of analysis. If these assumptions are not satisfied then appropriate shifting and scaling need to be performed. Since signal $x(t)$ is real and centered at $t = 0$, it will be centered around $\omega = 0$, which means $m_t = 0$ and $m_\omega = 0$.

Now from Cauchy-Schwarz inequality, we have following:

$$\left| \int_{-\infty}^{\infty} tx(t)\frac{d\,x(t)}{dt}dt \right|^2 \leq \int_{-\infty}^{\infty} |tx(t)|^2 dt \int_{-\infty}^{\infty} \left|\frac{dx(t)}{dt}\right|^2 dt \qquad (3.58)$$

By applying Parseval's theorem on the right-hand side (RHS) of Eq. (3.58), we have,

$$RHS = \sigma_t^2 \int_{-\infty}^{\infty} |j\omega X(\omega)|^2 d\omega = \sigma_t^2 \int_{-\infty}^{\infty} \omega^2 |X(\omega)|^2 d\omega$$

$$= \sigma_t^2 \sigma_\omega^2$$

and

$$\int_{-\infty}^{\infty} tx(t)\frac{dx(t)}{dt}dt = \frac{1}{2}\int_{-\infty}^{\infty} t\frac{dx^2(\tau)}{d\tau}dt$$

$$= \frac{1}{2}tx^2(t)|_{-\infty}^{\infty} - \frac{1}{2}\int_{-\infty}^{\infty} x^2(t)dt \qquad (3.59)$$

$$= -\frac{1}{2}$$

So, the left hand side (LHS) of Eq. (3.58) can be simplified as,

$$LHS = \left| \int_{-\infty}^{\infty} tx(t)\frac{dx(t)}{dt}dt \right|^2 = \frac{1}{4}$$

As $x(t) \in L^2(\mathbb{R})$, it decays fast so at limit $\infty$ first term becomes 0 and second term $\int_{-\infty}^{\infty} x^2(t)dt$ is 1 in Eq. (3.59).

Hence,

$$\frac{1}{4} \le \sigma_t^2 \sigma_\omega^2$$

$$\text{or} \quad \sigma_t^2 \sigma_\omega^2 \ge \frac{1}{4} \tag{3.60}$$

The Cauchy-Schwarz inequality becomes equality when two functions are collinear, which means one function can be expressed as a scalar multiplication of other function. It should be noted that only Gaussian kind of functions can follow this property. Hence, $x(t) = \alpha e^{-\beta t^2}$ (Gaussian function) holds the equality relation in this principle. This principle says that we can not make $\sigma_t^2$ or duration $(2\sigma_t)$ and $\sigma_\omega^2$ or bandwidth $(2\sigma_\omega)$ arbitrary small as they are bounded by some fixed number. In other words, we can not have such signal where both duration and bandwidth are very small, i.e., nearly zero. If we increase one quantity, for example duration, then according to this principle, bandwidth will be reduced and vice-versa.

## 3.8  INSTANTANEOUS FREQUENCY

Given an analytic signal $s(t) = s_R(t) + js_I(t)$, its AE functions $A(t)$ and instantaneous phase $\phi(t)$ are computed as [41, 42],

$$A(t) = \sqrt{s_R^2(t) + s_I^2(t)} \quad \text{and} \quad \phi(t) = \tan^{-1}\left(\frac{s_I(t)}{s_R(t)}\right) \tag{3.61}$$

The IF of the signal $s(t)$ can be calculated using Eq. (3.39) as,

$$\omega_i(t) = \frac{d \tan^{-1}\left[\frac{s_I(t)}{s_R(t)}\right]}{dt}$$

Further, it can be simplified as follows:

$$\omega_i(t) = \frac{1}{1 + \left(\frac{s_I(t)}{s_R(t)}\right)^2} \times \frac{s_R(t)s_I'(t) - s_I(t)s_R'(t)}{s_R^2(t)}$$

$$= \frac{s_R(t)s_I'(t) - s_I(t)s_R'(t)}{s_R^2(t) + s_I^2(t)}$$

$$= \frac{s_R(t)s_I'(t) - s_I(t)s_R'(t)}{A^2(t)}$$

$$\omega_i(t) = \frac{1}{A^2(t)}[s_R(t)s_I'(t) - s_I(t)s_R'(t)] \tag{3.62}$$

It should be noted that for a real signal $s(t)$, the spectrum satisfies the conjugate symmetry property, which means $S(-\omega) = S^*(\omega)$; due to this property, energy density spectrum $|S(\omega)|^2$ is always symmetric.

Due to this symmetric nature of the energy spectrum density, the computed average frequency will always be zero which does not make any sense. Therefore, we remove one part of the spectrum and consider the signal corresponding to positive frequencies which is known as analytic signal. The bandwidth and average frequency (mean frequency) computed from this signal are meaningful and make physical sense. The average frequency $(m_\omega)$ is given by,

$$\text{Average frequency: } m_\omega = \frac{1}{E} \int_0^\infty \omega |S(\omega)|^2 d\omega$$

Suppose $z(t)$ is an analytic signal corresponding to real signal $s(t)$, then

$$m_\omega = \frac{1}{E} \int_{-\infty}^\infty \omega |Z(\omega)|^2 d\omega$$

$$\text{The analytic signal: } z(t) = 2 \times \frac{1}{2\pi} \int_0^\infty S(\omega) e^{j\omega t} d\omega$$

The $z(t)$ can be considered as a general form of real signal $s(t)$. Due to this reason, a factor of 2 is used so that the real part of the analytic signal will be actual real signal. The spectrum $S(\omega)$ of the real signal $s(t)$ can be defined using Fourier transform as,

$$S(\omega) = \int_{-\infty}^\infty s(t) e^{-j\omega t} dt$$

$z(t)$ can be written in terms of inverse Fourier transform $S(\omega)$ as follows:

$$z(t) = \frac{1}{\pi} \int_0^\infty S(\omega) e^{j\omega t} d\omega$$

$$= \frac{1}{\pi} \int_0^\infty \int_{-\infty}^\infty s(t') e^{-j\omega t'} dt' \, e^{j\omega t} d\omega$$

$$= \frac{1}{\pi} \int_0^\infty \int_{-\infty}^\infty s(t') e^{j\omega(t-t')} dt' \, d\omega$$

It should be noted that,

$$\int_0^\infty e^{j\omega t} d\omega = \pi\delta(t) + \frac{j}{t}$$

Hence,

$$z(t) = \frac{1}{\pi} \int_{-\infty}^\infty \left[ s(t')\pi\delta(t-t') + \frac{js(t')}{(t-t')} \right] dt'$$

$$= \frac{1}{\pi} \int_{-\infty}^\infty s(t')\pi\delta(t-t') dt' + \frac{1}{\pi} \int_{-\infty}^\infty \frac{js(t')}{(t-t')} dt'$$

$$= s(t) + j \int_{-\infty}^\infty \frac{s(t')}{\pi(t-t')} dt' \tag{3.63}$$

This equation can be written as,

$$z(t) = s(t) + j \text{ [Hilbert transform of signal } s(t)]$$

$$z(t) = s(t) + j\, \mathrm{H}[s(t)] \tag{3.64}$$

where $\mathrm{H}[\cdot]$ denotes the Hilbert transform operation. The AE of the signal is determined as,

$$|z(t)| = \sqrt{s^2(t) + \mathrm{H}^2[s(t)]} \tag{3.65}$$

The instantaneous phase $\phi(t)$ is obtained as,

$$\phi(t) = \tan^{-1}\left[\frac{\mathrm{H}[s(t)]}{s(t)}\right] \tag{3.66}$$

The computation of IF can be carried out as the derivative of phase $\phi(t)$ as shown in Eq. (3.39).

The process of determining AE and IF functions using the analytic representation of signal in Eqs. (3.65) and (3.39), respectively is termed as the Hilbert transform separation algorithm (HTSA) [43]. For example, an AFM signal $x(n)$ represented in Eq. (3.20) has been considered for computing AE and IF functions using HTSA technique which is shown in Fig. 3.8.

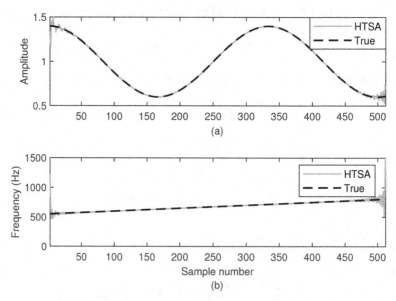

**Figure 3.8**   (a) AE function and (b) IF function computed using HTSA.

Hilbert transform of $s(t)$ is given by,

$$\mathrm{H}[s(t)] = \int_{-\infty}^{\infty} \frac{s(t')}{\pi(t - t')} dt' \tag{3.67}$$

or

$$H[s(t)] = s(t) * \frac{1}{\pi t} \qquad (3.68)$$

where $*$ denotes the convolution operator. Applying Fourier transform both sides, we have,

$$S_H(\omega) = -j \operatorname{sgn}(\omega) S(\omega) \qquad (3.69)$$

$$H(\omega) = -j \operatorname{sgn}(\omega) \qquad (3.70)$$

where $\operatorname{sgn}(\omega)$ is defined as,

$$\operatorname{sgn}(\omega) = \begin{cases} -1, & \omega < 0 \\ 0, & \omega = 0 \\ 1, & \omega > 0 \end{cases} \qquad (3.71)$$

$H(\omega)$ in Eq. (3.70) is frequency response of Hilbert transform operator. It can also be written as,

$$H(\omega) = \begin{cases} -j, & \omega > 0 \\ j, & \omega < 0. \end{cases} \qquad (3.72)$$

or

$$H(\omega) = \begin{cases} e^{-j\pi/2}, & \omega > 0 \\ e^{j\pi/2}, & \omega < 0 \end{cases} \qquad (3.73)$$

In the Hilbert transformation, magnitude of the frequency components present in the signal is same. The phase of negative frequency component of the signal is shifted by $+\pi/2$ and for positive frequency component, the phase is shifted by $-\pi/2$ [16]. The process of representing the signal energy into time-frequency domain using obtained AE and IF functions is known as Hilbert spectral analysis (HSA). The mathematical expression for TFR $H_s(t, \omega)$ of the signal $s(t)$ can be given as,

$$H_s(t, \omega) = \begin{cases} A^2(t); & \omega = \omega_i(t) \\ 0; & \omega \neq \omega_i(t) \end{cases} \qquad (3.74)$$

## 3.9 BASIC IDEAS RELATED TO TFDS

We have seen that the time-domain provides information about the variation of the signal with respect to time, and frequency-domain gives information about the different frequency components present in the signal. But individually they are not able to provide time information about the spectral changes in the signal. Time-frequency distribution (TFD) provides such analysis where time-varying frequency characteristics can be studied. TFD represents the energy density of the signal simultaneously in time and frequency-domains, which can be used in the similar way like any other density functions [39, 44, 45]. Assume for the signal $x(t)$, the energy density of the signal is $|x(t)|^2$ and spectral energy density of the signal is $|X(\omega)|^2$. Suppose, the energy intensity in time-frequency domain is denoted by $E(t, \omega)$. Table 3.2 lists some useful properties of ideal TFD.

## Table 3.2
## Properties of TFD

| Property | Mathematical Explanation |
|---|---|
| Marginals | Time marginal condition: $\int E(t,\omega)d\omega = \|x(t)\|^2$ <br> Frequency marginal condition: $\int E(t,\omega)dt = \|X(\omega)\|^2$ |
| Total energy | $E = \underbrace{\int\int E(t,\omega)dt\,d\omega}_{\text{Energy in time-frequency domain}} = \underbrace{\int \|x(t)\|^2 dt}_{\text{Energy in time-domain}}$ <br><br> $= \underbrace{\int \|X(\omega)\|^2 d\omega}_{\text{Energy in frequency-domain}}$ |
| Characteristic function | $C(\theta,\tau) = \int\int E(t,\omega)e^{j\theta t + j\tau\omega}dt\,d\omega$ |
| Time and frequency shifts | If $x(t) \to x(t-t_0)$ then $E(t,\omega) \to E(t-t_0,\omega)$ <br> If $X(\omega) \to X(\omega-\omega_0)$ then $E(t,\omega) \to E(t,\omega-\omega_0)$ <br> If $x(t) \to e^{j\omega_0 t}x(t-t_0)$ then $E(t,\omega) \to E(t-t_0,\omega-\omega_0)$ |
| Linear scaling | If $x(t) \to \sqrt{a}x(at)$, then $E(t,\omega) \to E(at, \frac{\omega}{a})$ <br><br> The modified distribution satisfies the following marginal conditions for scaled signal, $\sqrt{a}x(at)$ and its Fourier spectrum, $\frac{1}{\sqrt{a}}X(\omega/a)$: <br> $\int E(at, \frac{\omega}{a})d\omega = a\|x(at)\|^2$ <br> $\int E(at, \frac{\omega}{a})dt = \frac{1}{a}\|X(\frac{\omega}{a})\|^2$ |
| Weak and strong finite supports | Weak finite support: <br> If $x(t)=0$ for outside $(t_1,t_2)$, then <br> $E(t,\omega)=0$ for outside $(t_1,t_2)$. <br> Similarly, If $X(\omega)=0$ for outside $(\omega_1,\omega_2)$, then <br> $E(t,\omega)=0$ for outside $(\omega_1,\omega_2)$. <br><br> Strong finite support: <br> If $x(t)=0$ for a particular time, then $E(t,\omega)=0$ <br> If $X(\omega)=0$ for a particular frequency, then $E(t,\omega)=0$ |
| Uncertainty principle | Suppose $x(t) \leftrightarrow X(\omega)$ <br><br> For time- and frequency-domain separately, <br> $\sigma_t^2 = \frac{1}{E}\int(t-m_t)^2\|x(t)\|^2 dt$ <br> $\sigma_\omega^2 = \frac{1}{E}\int(\omega-m_\omega)^2\|X(\omega)\|^2 d\omega$. <br><br> For joint time-frequency domain, <br> $\sigma_t^2 = \frac{1}{E}\int\int(t-m_t)^2 E(t,\omega)dt\,d\omega$ <br> $= \frac{1}{E}\int(t-m_t)^2 E(t)dt$ <br> $\sigma_\omega^2 = \frac{1}{E}\int\int(\omega-m_\omega)^2 E(t,\omega)dt\,d\omega$ <br> $= \frac{1}{E}\int(\omega-m_\omega)^2 E(\omega)d\omega$. <br><br> When a TFD satisfies the marginal properties then it follows the uncertainty principle. |
| IF estimation | $\omega_i(t) = \dfrac{\int_{-\infty}^{\infty} \omega E(t,\omega)d\omega}{\int_{-\infty}^{\infty} E(t,\omega)d\omega}$ |

## PROBLEMS

Q 3.1 A signal $\phi(t)$ has mean time ($t_0$), time duration ($2\sigma_t$), mean frequency ($\omega_0$), and bandwidth ($2\sigma_\omega$). The signal is scaled by $a$ and shifted by $b$ and hence we get $\psi(t) = \frac{1}{\sqrt{a}}\phi\left(\frac{t-b}{a}\right)$. Compute the following for $a = 1$, $a = 0.5$, and $a = 2$ keeping $b = 5$:

(a) Mean time and time duration of the signal $\psi(t)$.

(b) Mean frequency and bandwidth of the signal $\psi(t)$.

Sketch the Heisenberg box of the signal $\psi(t)$ for all three cases and explain the effects of scaling and shifting on Heisenberg box.

Q 3.2 For signal $x(t) = e^{-\sigma t^2}e^{j\omega_0 t}$, compute the following parameters:

(a) Bandwidth due to AM $(B_{AM})$

(b) Bandwidth due to FM $(B_{FM})$

(c) Total bandwidth $(B_T)$

(d) Center frequency $(m_\omega)$

Q 3.3 Consider an AFM signal $x(t)$ defined as,

$$x(t) = A[1 + \mu\cos(\omega_m t)]\cos[\omega_c t + \beta\cos(\omega_m t)]$$

where $A = 1.4$, $\mu = 0.6$, $\omega_m = 80\pi$, $\omega_c = 2000\pi$, and $\beta = 0.72$. Determine the AE and IF functions using DESA-1, DESA-2, and HTSA techniques. Also compare them in terms of mean squared error (MSE) with respect to actual AE and IF functions in order to find out the best among the three methods. Use suitable sampling frequency for simulation.

Q 3.4 Consider a time-domain Gaussian function $x(t)$ expressed as,

$$x(t) = e^{-\sigma t^2}$$

Compute the 3 dB duration and 3 dB bandwidth of the signal $x(t)$ for three values of Gaussian parameter $\sigma = \sigma_1$, $\sigma_1/2$, $2\sigma_1$. Plot the Heisenberg uncertainty box and compute bandwidth-duration product for all three cases. Also comment on Heisenberg uncertainty principle based on the obtained results.

Q 3.5 Determine the analytic signal representation of the following signal models with condition $0 < \omega_2 < \omega_1$:

(a) $\cos(\omega_1 t) + \sin(\omega_2 t)$

(b) $\sin(\omega_1 t) + \cos(\omega_2 t)$

(c) $\cos(\omega_1 t) + \cos(\omega_2 t)$

(d) $\sin(\omega_1 t) + \sin(\omega_2 t)$

Q 3.6 Determine the energy of the following signals:

(a) $x(t) = \frac{\sin(5\pi t)}{\pi t}$

(b) $X(\omega) = \frac{\sin(\pi\omega/2)}{2\omega}$

(c) $x(t) = e^{-\sigma t + j\cos(42\pi t)} u(t)$

(d) $x(t) = e^{-\sigma t^2 + j\cos(90\pi t)}$

**Q 3.7** Consider a multi-component signal $x(t)$, defined as,

$$x(t) = A_1 \sin(\omega_1 t) + A_2 \sin(\omega_2 t)$$

Find out the expression for AE and IF functions for this signal using HTSA. In real-time scenario, when data is received in burst mode, which one from DESA and HTSA is preferable for computation of AE and IF functions of the signal?

**Q 3.8** Consider a multi-component signal $x(t)$, expressed as,

$$x(t) = \sum_{i=1}^{M} a_i(t) \cos[\phi_i(t)]$$

Perform the analysis of this signal using Gabor filter and DESA-1 algorithm. Also, perform synthesis of the signal using the obtained AE and IF functions from analysis step and compute the MSE of synthesized signal components with original signal components. Number of signal components ($M$) and their parameters $a_i(t)$ and $\phi_i(t)$ can be selected as per your choice. Also, consider suitable sampling rate for simulation.

**Q 3.9** Consider an AFM signal $x(t)$, expressed as,

$$x(t) = e^{-\alpha t^2 + j\beta t^2}$$

Compute the bandwidths of the signal due to AM and FM parts. Also determine the total bandwidth of the signal.

**Q 3.10** Consider the AFM signal $x(t)$ mentioned in Q 3.9. Transform $x(t)$ into spectral domain and then compute the duration due to SAM and SPM. Also compute the total duration of the signal.

**Q 3.11** Consider a rectangular pulse signal $s(t)$ defined by Eq. (3.29) and compute two Heisenberg box parameters mean time $m_t$ and duration $2\sigma_t$ for the signals $s(t - t_0)$, $s(at)$, and $s(t)e^{j\omega_0 t}$.

**Q 3.12** Compute the Heisenberg uncertainty box parameters, i.e., $m_t^{new}$, $2\sigma_t^{new}$, $m_\omega^{new}$, and $2\sigma_\omega^{new}$ of the modulated signal in time-domain, $x(t)e^{-j\omega_0 t}$ and frequency-domain, $X(\omega)e^{j\omega t_0}$. Given that the Heisenberg uncertainty box parameters for signal $x(t)$ are $m_t$, $2\sigma_t$, $m_\omega$, and $2\sigma_\omega$.

# 4 Short-Time Fourier Transform

*"The future belongs to those who believe in the beauty of their dreams."* –Eleanor Roosevelt

## 4.1 STFT

In conventional Fourier transform, time-domain and frequency-domain are disjoint which means time-domain of the signal does not have any information about the frequency-domain and frequency-domain does not have any information about the time-domain. So, combining the time-domain together with frequency-domain is the motivation for time-frequency analysis (TFA). The short-time Fourier transform (STFT) is a basic TFA technique where we multiply signal with the window function of finite duration and then we compute the Fourier transform of the windowed signal. We shift window function to other location and perform the same operation, i.e., Fourier transform of the windowed signal corresponding to that location. In the same way, we perform analysis of entire signal using windowing approach, the signal under analysis is assumed to be stationary signal. It should be noted that Fourier transform provides a global analysis of signal whereas the STFT provides local analysis and suitable for non-stationary signal analysis. Fourier transform converts 1D time-domain signal to 1D frequency-domain. On the other hand, STFT converts 1D time-domain signal to 2D time-frequency domain. Fourier transform provides frequency components present in the signal, whereas STFT provides IFs of the components present in the signal.

The mathematical expression for STFT is given as,

$$S(\tau, \omega) = \int_{-\infty}^{\infty} s(t)p(t-\tau)e^{-j\omega t} dt \qquad (4.1)$$

The signal $s(t)$ is multiplied with a window function $p(t-\tau)$, which is also known as analysis window, that is time-limited in nature and suppresses the signal outside a certain time-region where window is non-zero. In this way, Fourier transform gives a local spectrum of the signal corresponding to the particular window, parameter $\tau$ is used to shift the window at other location for computing the local spectrum for different window location. In this way, the entire signal is covered by varying shifting parameter $\tau$ of the window and all local spectrums are arranged together to obtain the STFT. The process of obtaining STFT is shown in Fig. 4.1.

The STFT $S(\tau, \omega)$ is a complex quantity and for representation purpose, we generally use squared magnitude of STFT which is known as spectrogram. The

DOI: 10.1201/9781003367987-4

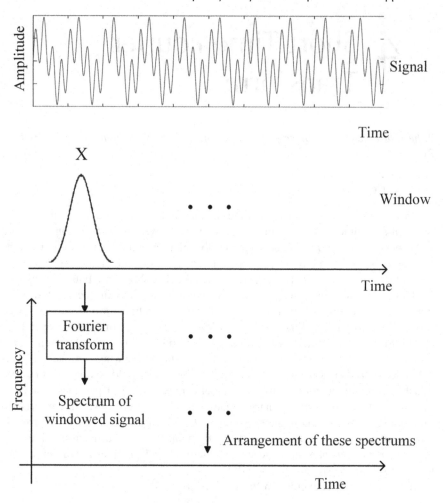

**Figure 4.1**  STFT-based analysis.

spectrogram of the signal, $s(t)$ can be defined as,

$$\text{Spectrogram} = |S(\tau, \omega)|^2 = \left| \int_{-\infty}^{\infty} s(t)p(t-\tau)e^{-j\omega t} dt \right|^2 \qquad (4.2)$$

Spectrogram shows the energy distribution of the signal in time-frequency domain where time is represented by $\tau$ parameter (window center) and frequency is represented by $\omega$ corresponding to exponential $e^{j\omega t}$. If $p(t)$ is chosen to be the Gaussian window, then the STFT is known as Gabor transform. An example of spectrogram for a linear chirp signal $x(n)$, mathematically defined by

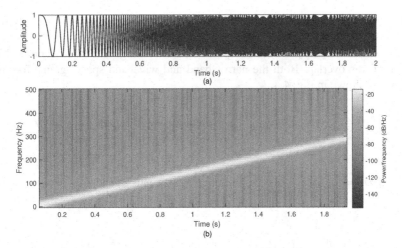

**Figure 4.2** (a) Signal $x(n)$ and (b) its spectrogram.

Eq. (4.3), is shown in Fig. 4.2.

$$x(n) = \cos\left[150\pi\left(\frac{n}{f_s}\right)^2\right], \quad \text{where } f_s = 1000 \text{ Hz} \qquad (4.3)$$

For plotting spectrogram, Hamming window of length 128 ms and 80% overlap is considered. The frequency of the analyzed signal is changing linearly with time which is not clear from time-domain representation of the signal, shown in Fig. 4.2 (a). On the other hand, the obtained TFR based on spectrogram clearly shows linear variation of the frequency for the considered signal. Based on the length of the window selected for the analysis of signal, we can have two different types of spectrogram, i.e., narrowband and wideband.

### 4.1.1 NARROWBAND AND WIDEBAND SPECTROGRAMS

The narrowband spectrogram provides a good frequency resolution in TFR, i.e., better spectral resolution of signal having sinusoids with closely spaced frequencies. In such analysis, generally the window size greater than two fundamental periods is considered. The wideband spectrogram provides a good temporal resolution in TFR, i.e., better temporal resolution of signals having impulses closely spaced in time [39, 46]. In wideband analysis, the window size less than a fundamental period is considered. For example, two signals of 500 ms duration, the first signal $x_1(n)$ with harmonics of 200 Hz sinusoidal mathematically defined by Eq. (4.4) and the other one $x_2(n)$ is periodic impulses with the time period of 20 ms mathematically defined by Eq. (4.5) have been considered. For spectrogram analysis of $x_1(n)$, the Hamming window of duration 25 ms with 20 ms overlap and 5 ms with 4 ms overlap for narrowband and wideband, respectively have been considered, and their results are shown in Fig. 4.3. The narrowband spectrogram of signal $x_2(n)$ is obtained using a Hamming window of duration 40 ms with 32 ms overlap. Similarly, the wideband spectrogram of $x_2(n)$ has been computed with Hamming window of duration 12 ms with 9.6 ms overlap. Both the narrowband and wideband spectrograms for signal $x_2(n)$ are shown in Fig. 4.4.

$$x_1(n) = \sum_{k=1}^{10} \cos\left(2\pi k\omega_0 \frac{n}{f_s}\right), \quad \text{where } \omega_0 = 200 \text{ Hz and } f_s = 5 \text{ kHz} \qquad (4.4)$$

$$x_2(n) = \begin{cases} 1, & \forall\, n = Pm \\ 0, & \text{otherwise} \end{cases} \quad \text{where } P = 100,\, m \in \mathbb{Z}, \text{ and } f_s = 5 \text{ kHz} \qquad (4.5)$$

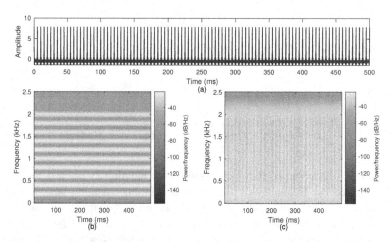

**Figure 4.3** (a) A signal with harmonics of 200 Hz and (b)–(c) its corresponding narrowband and wideband spectrograms, respectively.

Speech analysis greatly utilizes the concept of narrowband and wideband spectrograms due to the harmonic structure of voiced speech signal. In order to obtain narrowband spectrogram for voiced speech signal analysis, we use a longer duration window function typically more than two pitch periods. Under this condition, it has been observed that the main lobes of windows are not overlapped and hence letting the pitch harmonics visible in spectrogram. Whereas, analysis of voiced speech signal using wideband spectrogram requires a smaller duration window typically less than or equal to one pitch period. The overlapping of main lobes at pitch harmonics smears the harmonics line structure in spectrogram and preserves the resonance structure of vocal tract, i.e., formant of voiced speech signal. The simulation results for speech signal analysis using narrowband and wideband spectrograms are shown in Fig. 4.5. The detail of the speech signal can be found in section 1.1. For narrowband spectrogram, a 25 ms Hamming window with 90% overlap is used, whereas a 5 ms Hamming window with 90% overlap is used to obtain wideband spectrogram.

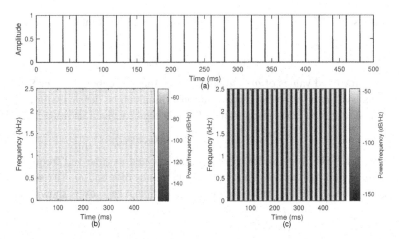

**Figure 4.4** (a) A signal with periodic impulses of period 0.02 seconds and (b)–(c) its corresponding narrowband and wideband spectrograms, respectively.

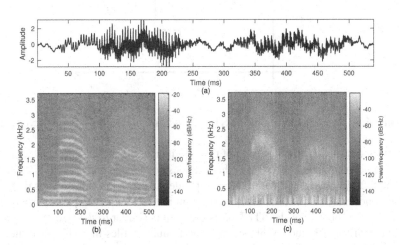

**Figure 4.5** (a) Speech signal and its corresponding (b) narrowband spectrogram and (c) wideband spectrogram.

## 4.2  TIME-FREQUENCY RESOLUTION OF STFT

The basis functions in STFT can be given by,

$$\phi_{\tau,\omega}(t) = p(t - \tau)e^{j\omega t} \tag{4.6}$$

The STFT basis functions expressed by Eq. (4.6) can be obtained by applying the time-shifting and frequency-shifting (or modulation) properties on window function $p(t)$.

Suppose, the Heisenberg box representation for window funtion $p(t)$ is expressed as,

$$[(m_t, m_\omega); (2\sigma_t, 2\sigma_\omega)] \tag{4.7}$$

Then, for fixed values of $\omega = \omega_0$ and $\tau = \tau_0$, the Heisenberg box for STFT basis functions will have following representation, which can be derived with help of the properties of scaling and modulation on Heisenberg box given in Table 3.1:

$$[(m_t + \tau_0, m_\omega + \omega_0); (2\sigma_t, 2\sigma_\omega)] \tag{4.8}$$

Now, the STFT basis functions can be interpreted as time and frequency shift-ings of the window function $p(t)$ for different values of $\tau$ and $\omega$. The sketch of the Heisenberg box for STFT basis function for $\tau = \tau_0$ and $\omega = \omega_0$ is shown in Fig 4.6.

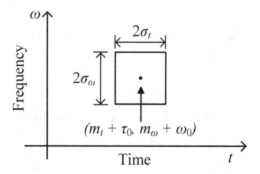

**Figure 4.6**   Representation of the STFT basis function in time-frequency plane.

It should be clear that location of time-frequency tile or Heisenberg box depends on parameters $\tau$ and $\omega$. The size of the box does not depend on parameters $\tau$ and $\omega$ as $2\sigma_t$ and $2\sigma_\omega$ (length and width) are independent of $\tau$ and $\omega$. This means that STFT basis functions provide an uniform resolution in time-frequency plane for any value of $\tau$ and $\omega$. The size of the STFT basis function is similar to the size of the window function in time-frequency plane. All time-frequency tiles in STFT have the same shape. The use of wide window requires for good frequency resolution but it will provide poor time resolution, on the other hand, narrow analysis window provides good time resolution and poor frequency resolution. Uncertainty principle provides a lower bound on the area of window. The joint time-frequency resolution of the STFT basis function is proportional to $\frac{1}{\sigma_t \sigma_\omega}$. The value $4\sigma_t \sigma_\omega$ represents the area of time-frequency tile. It should be noted that Gaussian window provides minimum value of $\sigma_t \sigma_\omega$ or minimum area of time-frequency tile and hence, provides highest time-frequency resolution. If we decrease $\sigma_\omega$, then $\sigma_t$ will increase in order to maintain the constant area of time-frequency tile.

## 4.3   STFT INTERPRETATIONS

The STFT can be interpreted in so many ways to perform signal analysis as follows:

## 4.3.1 FOURIER SPECTRUM OF THE WINDOWED SIGNAL

The STFT $S(\tau, \omega)$ can be considered as a Fourier transform of the windowed or time-limited signal, $s_p(t) = s(t)p(t - \tau)$ [47]. From Eq. (4.1), the STFT of signal $s(t)$ can be expressed as follows:

$$S(\tau, \omega) = \int_{-\infty}^{\infty} s_p(t)e^{-j\omega t} dt \qquad (4.9)$$

The STFT provides frequency contents of the windowed signal at different locations which are determined by value of $\tau$ and it provides local analysis.

## 4.3.2 STFT AS INNER PRODUCT

The STFT of signal $s(t)$, given in Eq. (4.1), can be expressed as,

$$S(\tau, \omega) = \int_{-\infty}^{\infty} s(t) \left[ p(t - \tau)e^{-j\omega t} \right] dt \qquad (4.10)$$

Which can be written in the form of inner product between signal $s(t)$ and modulated window $p(t - \tau)e^{j\omega t}$. The window $p(t)$ is considered to be a real valued function.

$$S(\tau, \omega) = < s(t),\ p(t - \tau)e^{j\omega t} > \qquad (4.11)$$

This means that the STFT measures the similarity between signal $s(t)$ and modulated window $p(t - \tau)e^{j\omega t}$.

Using Parseval's theorem, the inner product of the signal $s(t)$ and the modulated window $p(t - \tau)e^{j\omega t}$ can be considered as a similarity measurement between the Fourier spectrum of $s(t)$ and Fourier spectrum of $p(t - \tau)e^{j\omega t}$. Such concept helps to implement STFT using filtering in frequency-domain.

## 4.3.3 STFT AS CONVOLUTION

The STFT of signal $s(t)$ which is expressed in Eq. (4.1) can be written as,

$$S(\tau, \omega) = \int_{-\infty}^{\infty} s(t)e^{-j\omega t} p(t - \tau) dt \qquad (4.12)$$

From, the above expression, STFT can be written in the convolution form as,

$$S(\tau, \omega) = s(\tau)e^{-j\omega \tau} * p(-\tau) \qquad (4.13)$$

Figure 4.7 shows the representation of STFT in convolution form. The convolution form of the STFT explains the filtering process for a fixed value of frequency $\omega$ with respect to time.

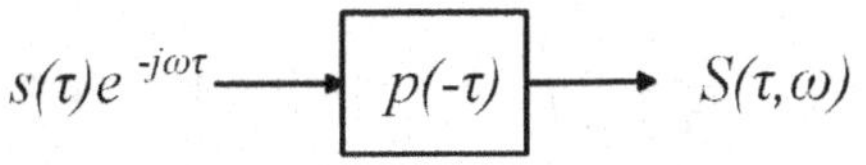

**Figure 4.7** Convolution form representation of STFT.

### 4.3.4   IMAGE AND CONTOUR FORMS

It should be noted that STFT is a complex quantity, and squared magnitude of the STFT, which is known as a spectrogram, is used for TFR. The spectrogram maps 1D signal into 2D TFR where energy is plotted as a function of time and frequency. Since energy is positive quantity and can be sketched in the form of image or also represented by contour plot which can trace the frequencies corresponding to significant energy as a function of time.

Figure 4.8. shows bat signal, spectrogram-based TFR, image and contour plots of the bat signal.

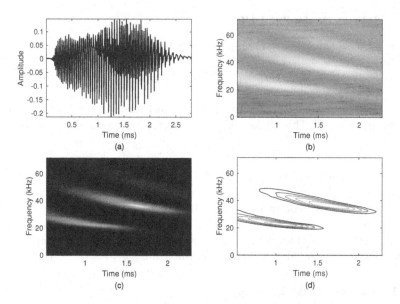

**Figure 4.8**   (a) Bat signal, (b) TFR of bat signal based on spectrogram, (c) image representation for the spectrogram of bat signal, and (d) contour plot of the spectrogram for bat signal. The author wishes to thank Curtis Condon, Ken White, and Al Feng of the Beckman Institute of the University of Illinois for the bat data and for permission to use it in this book.

### 4.4   RECONSTRUCTION PROCESS FOR STFT

The signal $s(t)$ can be reconstructed from its STFT, $S(\tau, \omega)$ [48]. The reconstruction process for STFT can be expressed mathematically as follows:

$$s(t) = \frac{1}{2\pi} \int_{-\infty}^{\infty} \int_{-\infty}^{\infty} S(\tau, \omega) q(t - \tau) e^{j\omega t} d\tau d\omega \qquad (4.14)$$

In the above-mentioned expression, $S(\tau, \omega)$ is defined by Eq. (4.1). The following condition needs to be satisfied in order to have synthesis of the signal:

$$\int_{-\infty}^{\infty} p(t)q^*(t)dt = 1 \tag{4.15}$$

Here, $p(t)$ and $q(t)$ are termed as analysis window and synthesis window, respectively. The above-mentioned condition in frequency-domain can be expressed as,

$$\frac{1}{2\pi}\int_{-\infty}^{\infty} P(\omega)Q^*(\omega)d\omega = 1 \tag{4.16}$$

Let us consider the RHS of the Eq. (4.14) in order to prove the reconstruction formula of STFT.

$$\text{RHS} = \frac{1}{2\pi}\int_{-\infty}^{\infty}\int_{-\infty}^{\infty} S(\tau, \omega)q(t-\tau)e^{j\omega t}d\tau d\omega \tag{4.17}$$

The convolutional form of the STFT $S(\tau, \omega)$, as given in Eq. (4.13), can be written in frequency-domain with the assumption $p(t) = p(-t)$ as,

$$S(\omega, \omega') = P(\omega')S(\omega + \omega') \tag{4.18}$$

The $q(t-\tau)$ in frequency-domain using shifting property of Fourier transform can be written as,

$$Q_s(\omega') = Q(\omega')e^{-j\omega' t}$$

Now, using Parseval theorem RHS can be written as,

$$\text{RHS} = \frac{1}{2\pi}\int_{-\infty}^{\infty}\frac{1}{2\pi}\int_{-\infty}^{\infty} S(\omega' + \omega)P(\omega')Q^*(\omega')e^{j\omega' t}e^{j\omega t}d\omega' d\omega \tag{4.19}$$

RHS can be written as,

$$\text{RHS} = \frac{1}{2\pi}\int_{-\infty}^{\infty} P(\omega')Q^*(\omega')\frac{1}{2\pi}\int_{-\infty}^{\infty} S(\omega' + \omega)e^{j(\omega+\omega')t}d\omega d\omega' \tag{4.20}$$

Using the synthesis expression of Fourier transform, RHS can be written as,

$$\text{RHS} = s(t)\frac{1}{2\pi}\int_{-\infty}^{\infty} P(\omega')Q^*(\omega')d\omega' \tag{4.21}$$

In case of RHS equals to $s(t)$, then following condition needs to be satisfied:

$$\frac{1}{2\pi}\int P(\omega')Q^*(\omega')d\omega' = 1 \tag{4.22}$$

In time-domain, the same condition can be written as,

$$\int_{-\infty}^{\infty} p(t)q^*(t)dt = 1 \qquad (4.23)$$

It should be noted that if the same window is used for analysis and synthesis, that is $p(t) = q(t)$, in that case, the required condition for reconstruction of the signal in time-domain is as,

$$\int_{-\infty}^{\infty} |p(t)|^2 dt = 1 \qquad (4.24)$$

and this required condition in frequency-domain can be expressed as,

$$\frac{1}{2\pi} \int |P(\omega')|^2 d\omega' = 1 \qquad (4.25)$$

It should be noted that signal reconstruction from STFT requires STFT for the entire time and frequency ranges with the above-mentioned conditions on the analysis and synthesis window functions.

## 4.5   ENERGY CONSERVATION FOR STFT

Parseval's theorem also works for STFT-based analysis. According to this energy conservation rule, the energy of the signal in time-domain is similar to the energy in STFT-based time-frequency domain. This Parseval's theorem for STFT can be expressed as,

$$\underbrace{\int_{-\infty}^{\infty} |s(t)|^2 \, dt}_{\text{Energy of the signal in time-domain}} = \underbrace{\frac{1}{2\pi} \int_{-\infty}^{\infty} \int_{-\infty}^{\infty} |S(\tau,\omega)|^2 \, d\tau d\omega}_{\text{Energy of the signal in time-frequency domain}} \qquad (4.26)$$

In order to prove the Parseval's theorem for STFT, consider the RHS of the Eq. (4.26).

$$\text{RHS} = \frac{1}{2\pi} \int_{-\infty}^{\infty} \left[ \int_{-\infty}^{\infty} |S(\tau,\omega)|^2 d\tau \right] d\omega \qquad (4.27)$$

which can be written using the frequency-domain expression for the convolutional form of STFT $S(\tau,\omega)$ with symmetric analysis window shown in Eq. (4.18) in the above expression as follows:

$$\text{RHS} = \frac{1}{2\pi} \int_{-\infty}^{\infty} \left[ \frac{1}{2\pi} \int_{-\infty}^{\infty} \left| S\left(\omega,\omega'\right) \right|^2 d\omega' \right] d\omega \qquad (4.28)$$

$$\text{RHS} = \frac{1}{2\pi} \int_{-\infty}^{\infty} \frac{1}{2\pi} \int_{-\infty}^{\infty} \left| P(\omega')S\left(\omega+\omega'\right) \right|^2 d\omega \, d\omega' \qquad (4.29)$$

By applying Parseval's theorem, the above expression can be written as,

$$\text{RHS} = E_s \frac{1}{2\pi} \int_{-\infty}^{\infty} |P(\omega')|^2 d\omega'$$

where $E_s$ is the energy of the signal $s(t)$. This expression can be interpreted as,

Energy in STFT domain = Energy of the signal × Energy of the window function

If, energy of the analysis window function is one, i.e.,

$$\frac{1}{2\pi} \int_{-\infty}^{\infty} |P(\omega)|^2 d\omega = \int_{-\infty}^{\infty} |p(t)|^2 dt = 1 \tag{4.30}$$

In that case, perfect reconstruction of the signal can be obtained. It should be noted that, total energy of the signal in STFT domain is given as the product of energy of the signal and energy of the window function. Use of analysis window with unit energy helps in STFT domain to follow the Parseval's theorem.

## 4.6   SHORT-FREQUENCY FOURIER TRANSFORM

In STFT, the segmentation of the signal $s(t)$ is carried out in time-domain using window $g(t)$. On the other hand, in short-frequency Fourier transform (SFFT), the segmentation of the spectrum is performed using filter in order to obtain TFR [49, 39]. The band-pass component of the signal $s(t)$, at frequency $\omega$, is obtained from the spectrum of the signal multiplied by the shifted spectrum of the filter by frequency $\omega$. After that by performing inverse Fourier transform we obtain TFR of the signal. The SFFT of signal $s(t)$ can be expressed mathematically as,

$$S_{\text{SFFT}}(\tau, \omega) = \frac{1}{2\pi} \int_{-\infty}^{\infty} S(\omega')G(\omega' - \omega)e^{j\omega'\tau}d\omega' \tag{4.31}$$

The square of the SFFT is termed as the sonograph which is given by,

$$\text{Sonograph} = |S_{\text{SFFT}}(\tau, \omega)|^2 \tag{4.32}$$

Selection of bandwidth of band-pass filter in sonograph affects the time-delay likewise window length affects the instantaneous frequency in the spectrogram.

SFFT can also be expressed in convolution form as,

$$S_{\text{SFFT}}(\tau, \omega) = s(\tau) * g(\tau)e^{j\omega\tau} \tag{4.33}$$

which can be written as,

$$S_{\text{SFFT}}(\tau, \omega) = e^{j\omega\tau} \int_{-\infty}^{\infty} s(t)g(\tau - t)e^{-j\omega t}dt \tag{4.34}$$

With the help of above expression, the sonograph can be expressed as,

$$\text{Sonograph} = \left| \int_{-\infty}^{\infty} s(t)g(\tau - t)e^{-j\omega t}dt \right|^2 \tag{4.35}$$

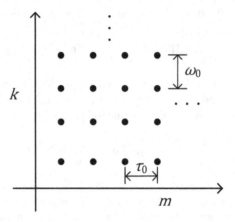

**Figure 4.9**   The sampling grid for discrete STFT.

By comparing Eq. (4.2) with Eq. (4.35) we have,

$$p(t) = g(-t) \tag{4.36}$$

In other words, when relation in Eq. (4.36) is satisfied, the spectrogram (Eq. (4.2)) and sonograph (Eq. (4.35)) based TFRs become similar and using the Eqs. (4.1) and (4.34), the STFT and SFFT are related as,

$$S_{\text{SFFT}}(\tau, \omega) = e^{j\omega\tau} S(\tau, \omega)$$

## 4.7   DISCRETE VERSION OF STFT

The discretization of $(\omega, \tau)$ parameters is required in order to obtain discrete STFT which can be expressed as, $\omega = k\omega_0$ and $\tau = m\tau_0$ [50, 51].

The discrete version of STFT can be expressed as,

$$S_d(m,k) = \int_{-\infty}^{\infty} s(t)p(t - m\tau_0)e^{-jk\omega_0 t} dt \tag{4.37}$$

After computing $S_d(m,k)$ for all values of $m$ and $k$, the interpolation process is carried out for interpolating these grid points in order to obtain $S_d(\omega, \tau)$. Figure 4.9 shows the sampling grid points for discrete STFT.

The signal $s(t)$ can be reconstructed using discrete version of STFT $S_d(m,k)$ if $\omega_0\tau_0 < 2\pi$ as,

$$s(t) = \frac{\omega_0\tau_0}{2\pi} \sum_k \sum_m S_d(m,k)e^{jk\omega_0 t} q(t - m\tau_0) \tag{4.38}$$

It should be noted that the discrete STFT-based transform domain is redundant and in order to reduce redundancy the sampling grid can be enlarged using the condition, $\omega_0 \tau_0 = 2\pi$. The set of basis functions follows orthogonality for this condition. On the other hand, these orthogonal basis functions are not well localized in time-domain or frequency-domain. Due to this reason, oversampling of STFT is preferred and this type of series expansion for STFT is also termed as Gabor expansion.

For a discrete-time signal $s(n)$, the analysis equation for discrete STFT (discrete-time and continuous frequency grids) is given as,

$$S(m, e^{j\omega}) = \sum_n s(n)p(n-mN)e^{-j\omega n} \tag{4.39}$$

We assume that sampling rate of the signal is higher by the factor of $N$. For practical applications, $\omega$ is also discretized as,

$$\omega_k = \frac{2\pi k}{M}, \quad k = 0, 1, \ldots, M-1$$

Now, analysis expression for discrete STFT is given by the following equation:

$$\text{Analysis expression: } S(m,k) = \sum_n s(n)p(n-mN)e^{\frac{-j2\pi nk}{M}} \tag{4.40}$$

The reconstructed signal $\hat{s}(n)$ can be computed by following mathematical expression:

$$\text{Synthesis expression: } \hat{s}(n) = \sum_{m=-\infty}^{\infty} \sum_{k=0}^{M-1} S(m,k)q(n-mN)e^{\frac{j2\pi nk}{M}} \tag{4.41}$$

The reconstruction process for discrete STFT can be proved as follows: For simple analysis, assume $N = 1$ (no sub-sampling) and $q(n) = \delta(n)$.

Synthesis expression as shown in Eq. (4.41), can be expressed as,

$$\hat{s}(n) = \sum_{m=-\infty}^{\infty} \sum_{k=0}^{M-1} S(m,k)\delta(n-m)e^{\frac{j2\pi nk}{M}}$$

$$= \sum_{k=0}^{N-1} S(n,k)e^{\frac{j2\pi nk}{M}} \tag{4.42}$$

Putting $N = 1$ and $m = n$ in analysis expression (Eq. (4.40)), we have,

$$S(n,k) = s(n)p(0)e^{\frac{-j2\pi nk}{M}} \tag{4.43}$$

Putting $S(n,k)$ in the Eq. (4.42), we have,

$$\hat{s}(n) = \sum_{k=0}^{M-1} s(n)p(0)$$

$$= s(n)p(0)M \tag{4.44}$$

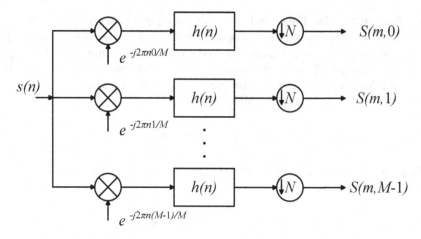

**Figure 4.10**  Analysis part of discrete STFT based on filter bank method.

In order to have proper reconstruction, $p(0) = \frac{1}{M}$ or $p(0)M = 1$. This reconstruction method for discrete STFT is termed as spectral summation method or filter bank summation method.

The discrete STFT can also be realized using filter bank method, which is described in the consequent sections [51]. The analysis equation of discrete STFT, given in Eq. (4.40), can be considered as filtering with impulse response $h(n) = p(-n)$ of $s(n)e^{\frac{-j2\pi nk}{M}}$ signal. Equation (4.41) can be interpreted as filtering the $S(m,k)$ with $q(n)$ followed by modulation. The analysis window $p(n)$ and synthesis window $q(n)$ generally have low-pass characteristics. The low-pass filter realization of the analysis and synthesis of the STFT are shown in Figs. 4.10 and 4.11.

The discrete STFT can also be realized using band-pass filters. The analysis equation for discrete STFT as given in Eq. (4.40) can be rewritten as,

$$S(m,k) = e^{\frac{-j2\pi kmN}{M}} \sum_n s(n)p(n-mN)e^{\frac{-j2\pi}{M}k(n-mN)} \tag{4.45}$$

This analysis process can be realized using band-pass filtering of the signal $s(n)$ with the following band-pass filters:

$$h_k(n) = p(-n)e^{j\frac{2\pi}{M}kn} \tag{4.46}$$

with the subsequent modulation process.

The synthesis equation of STFT as given in Eq. (4.41) can be written as,

$$\hat{s}(n) = \sum_{m=-\infty}^{\infty} \sum_{k=0}^{M-1} S(m,k)e^{j\frac{2\pi}{M}kmN}q(n-mN)e^{j\frac{2\pi}{M}k(n-mN)} \tag{4.47}$$

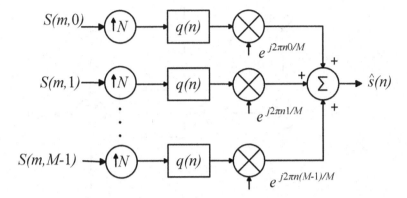

**Figure 4.11**  Synthesis part in the filtering representation of discrete STFT.

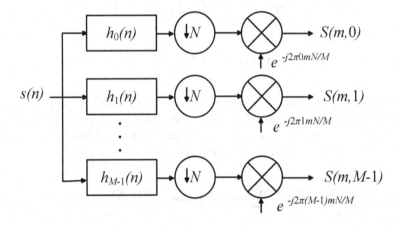

**Figure 4.12**  Analysis part of STFT using band-pass filtering.

The realization of the synthesis process requires modulation of the STFT and then filtering using the following band-pass filters:

$$q_k(n) = q(n)e^{j\frac{2\pi}{M}kn} \qquad (4.48)$$

Figures 4.12 and 4.13 show the realization of discrete STFT based on band-pass filtering method. The reconstruction of the signal from the STFT can be obtained in other way, based on the Fourier transform interpretation of STFT, namely, overlap add method. The STFT can be interpreted as the Fourier transform of the windowed signal as given in section 4.3.1, which can be mathematically expressed as,

$$s(n)p(m-n) = \frac{1}{M}\sum_{k=0}^{M-1} S(m,k)e^{j\frac{2\pi nk}{M}} \qquad (4.49)$$

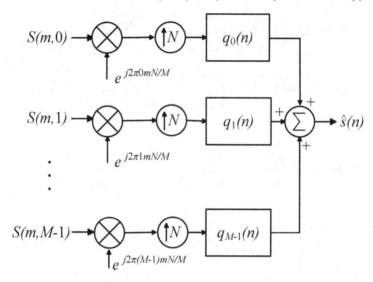

**Figure 4.13**   Realization of discrete STFT using band-pass filtering.

Samples of STFT are $B$ samples apart which can be written as $S(bB, k)$, where $b$ varies in the range $0 \leq b \leq M - 1$. Let us define the inverse DFT of $S(bB, k)$ as $z_b(n)$ which is equal to $s(n)p(bB - n)$ and it can be seen from Eq. (4.49). If we simply add all the inverse Fourier transform of STFT $z_b(n)$, we get,

$$z(n) = \sum_{b=-\infty}^{\infty} z_b(n) = \sum_{b=-\infty}^{\infty} s(n)p(bB - n) = s(n) \sum_{b=-\infty}^{\infty} p(bB - n) \qquad (4.50)$$

If $p(n)$ is a bandlimited signal and STFT is properly sampled by very small value $B$ to avoid time aliasing, then we can write,

$$\sum_{b=-\infty}^{\infty} p(bB - n) \approx \frac{P(0)}{B} \qquad (4.51)$$

where, $P(0)$ is the sum of all time-domain samples of the window $p(n)$. Using the above relation, Eq. (4.50) can be written as,

$$z(n) = s(n)\frac{P(0)}{B}$$

which shows that proper reconstruction of signal from STFT with a constant multi-plication factor $\frac{P(0)}{B}$, can be obtained by performing overlapping and add process for different window segments [52, 53].

Rectangular window with no overlap and Hamming windows with 50% overlap which satisfy the perfect reconstruction condition of STFT as given in Eq. (4.51) are shown in Fig. 4.14.

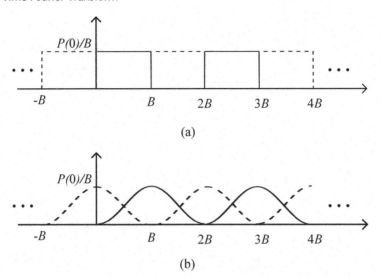

**Figure 4.14** Arrangement of window for reconstruction of signal from STFT using overlap add method: (a) rectangular window and (b) Hamming window.

## 4.8  EXAMPLES OF STFT

### 4.8.1  TIME SHIFTED SIGNAL

Consider $s_1(t)$ which is the shifted version of the signal $s(t)$ defined as,

$$s_1(t) = s(t - t_0) \tag{4.52}$$

Using Eq. (4.1), STFT of the signal $s_1(t)$ can be given by,

$$S_1(\tau, \omega) = \int_{-\infty}^{\infty} s(t - t_0)p(t - \tau)e^{-j\omega t}\,dt$$

By putting $t - t_0 = \alpha$ and $dt = d\alpha$, and further simplification provides,

$$S_1(\tau, \omega) = e^{-j\omega t_0} S(\tau - t_0, \omega) \tag{4.53}$$

It can be stated that, shifting of the signal introduces a factor of $e^{-j\omega t_0}$ and delay or shifting in STFT by the same amount. As an example, a complex-valued time-localized signal $x(n)$ is considered to show the effect of shift of signal on its spectrogram. The signal $x(n)$ can be mathematically represented as,

$$x(n) = w(n)\left[10e^{j\frac{5}{32}\pi n}\right] \tag{4.54}$$

with

$$w(n) = \begin{cases} w_H(n), & \text{if } 108 \le n \le 348 \\ 0, & \text{otherwise} \end{cases} \tag{4.55}$$

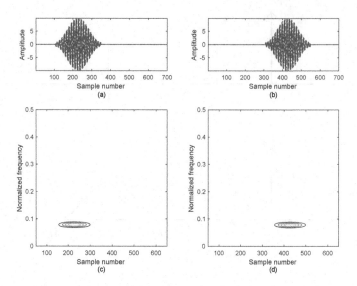

**Figure 4.15** (a) Real and imaginary parts of signal $x(n)$ are shown in solid and dash-dot lines, respectively, (b) real and imaginary parts of signal $x(n-200)$ are shown in solid and dash-dot lines, respectively. Their respective spectrograms are shown in (c) and (d).

where $w_H(n)$ is a Hamming window which localizes the signal in sample range [108–348], and signal $x(n)$ has length of $N = 700$ samples. The spectrogram of the signal $x(n)$ and shifted signal $x(n-200)$ is shown in Fig. 4.15.

### 4.8.2 FREQUENCY SHIFTED SIGNAL

Suppose frequency shifted or modulated signal $s_1(t)$ is expressed by,

$$s_1(t) = s(t)e^{j\omega_0 t} \tag{4.56}$$

Now, STFT of this signal $s_1(t)$ can be computed using Eq. (4.1) as,

$$S_1(\tau,\omega) = \int_{-\infty}^{\infty} s(t)e^{j\omega_0 t}p(t-\tau)e^{-j\omega t}dt$$

$$= \int_{-\infty}^{\infty} s(t)p(t-\tau)e^{-jt(\omega-\omega_0)}dt$$

$$= S(\tau,\omega-\omega_0) \tag{4.57}$$

Frequency shift property, shifts the frequency parameter of the STFT which can also be termed as time-domain modulation property. As an example, the signal $x(n)$ represented in Eq. (4.54) and its modulated version, $x(n)e^{j\frac{25}{64}\pi n}$ is considered. Their respective spectrograms are shown in Fig. 4.16.

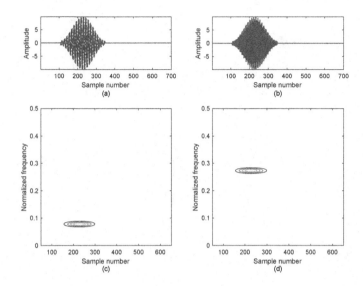

**Figure 4.16** (a) Real and imaginary parts of signal $x(n)$ are shown in solid and dash-dot lines, respectively, (b) real and imaginary parts of signal $x(n)e^{j\frac{25}{64}\pi n}$ are shown in solid and dash-dot lines, respectively. Their respective spectrograms are shown in (c) and (d).

### 4.8.3 TIME AND FREQUENCY SHIFTED SIGNAL

Suppose in this case, the modified signal is given by,

$$s_1(t) = s(t - t_0)e^{j\omega_0 t} \tag{4.58}$$

then its STFT can be expressed with the help of Eq. (4.1) as,

$$S_1(\tau, \omega) = \int_{-\infty}^{\infty} s(t - t_0)e^{j\omega_0 t} p(t - \tau)e^{-j\omega t} dt \tag{4.59}$$

$$= \int_{-\infty}^{\infty} s(t - t_0)p(t - \tau)e^{-jt(\omega - \omega_0)} dt \tag{4.60}$$

Now, put $t - t_0 = \alpha$ and $dt = d\alpha$, and after further simplifications, we have,

$$S_1(\tau, \omega) = \int_{-\infty}^{\infty} s(\alpha)p(\alpha - (\tau - t_0))e^{-j(\omega - \omega_0)\alpha}e^{-j(\omega - \omega_0)t_0} d\alpha$$

$$= e^{-j(\omega - \omega_0)t_0} S(\tau - t_0, \omega - \omega_0) \tag{4.61}$$

It can be interpreted from Eq. (4.61) that both time and frequency shifting operations together of the signal introduce a shift in STFT in time- and frequency-domain along with an additional multiplication factor, $e^{-j(\omega - \omega_0)t_0}$. As an example, the signal $x(n)$ represented in Eq. (4.54) and its shifted and modulated version, $x(n - 200)e^{j\frac{25}{64}\pi n}$ is considered. Their respective spectrograms are shown in Fig. 4.17.

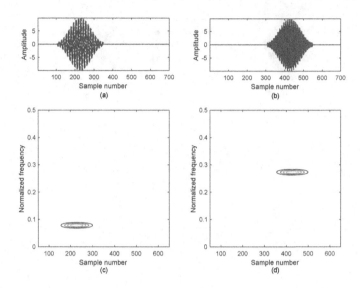

**Figure 4.17** (a) Real and imaginary parts of signal $x(n)$ are shown in solid and dash-dot lines, respectively, (b) real and imaginary parts of signal $x(n-200)e^{j\frac{25}{64}\pi n}$ are shown in solid and dash-dot lines, respectively. Their respective spectrograms are shown in (c) and (d).

## PROBLEMS

Q 4.1 Determine and draw the squared magnitude of STFT of the following signals using window, $w(t) = e^{-\alpha t^2}$:

(a) $x(t) = e^{j\omega_1 t}$

(b) $x(t) = \delta(t - \tau)$

Q 4.2 Consider the window $w(t)$ of duration $L$ for the analysis of signal using STFT technique at two extremes, i.e., $L = 0$ and $L \to \infty$ which can be mathematically represented by $w(t) = \delta(t)$ and constant function, $w(t) = c$, respectively. Discuss the characteristics of the STFT in both cases for an arbitrary signal $x(t)$.

Q 4.3 Consider a signal $x(t) = \cos(20\pi t) + \cos(200\pi t)$ sampled at sampling frequency $f_s = 600$ Hz. For a Gaussian window $w(t) = e^{\frac{-5t^2}{2}}$ of a fixed length $L$, study the nature of resolution for low-frequency as well as high-frequency components present in the signal $x(t)$.

Q 4.4 For the following signal models:

(a) $x_1(t) = \cos[2\pi(f_0 + f_1 t)t]$

(b) $x_2(t) = \cos[2\pi(f_2 - f_3 t)t]$

(c) $x_3(t) = \cos\left[2\pi(f_4 + f_5 t^2)t\right]$

(d) $x_4(t) = \cos\left[2\pi f_6 t + 25\cos(2\pi f_7 t)\right]$

The spectrograms of signals $x_1(t), x_2(t), x_3(t)$, and $x_4(t)$, are shown in Figs. 4.18 (a)–(d), respectively. Find out the parameters $f_i \, \forall \, i \in [1,7] \cap \mathbb{Z}$.

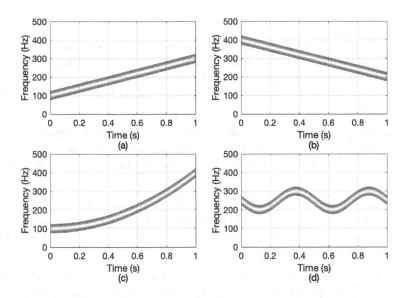

**Figure 4.18** (a)-(d) The spectrograms of signals $x_1(t)$, $x_2(t)$, $x_3(t)$, and $x_4(t)$ for Q 4.4.

Q 4.5 Generally, in speech signal processing, we prefer overlapped window for STFT computation in order to reduce the discontinuity and avoid loss of information. Verify the inversion of STFT using overlap add method based on simulation for any speech signal.

Q 4.6 Determine the squared magnitude of STFT of the following signals using window, $w(t) = e^{-\alpha t^2}$:

(a) $x(t) = e^{j\omega_1 t} + e^{j\omega_2 t}$

(b) $x(t) = \delta(t - t_1) + \delta(t - t_2)$

Also comment on the presence of cross-term in squared magnitude of STFT.

Q 4.7 Consider the signal $x(t) = \cos(\omega_m t)\cos(\omega_c t)$, where $\omega_c > \omega_m$. Perform Fourier analysis and STFT of this signal and also comment on how Fourier analysis result misleads the interpretation Choose the suitable values of $\omega_m$, $\omega_c$, and sampling frequency for simulation.

Q 4.8 Compute and plot the spectrogram of the following signal:

$$x(t) = \begin{cases} \sin(\omega_1 t), & \text{if } 0 \leq t < t_1 \\ \sin(\omega_2 t), & \text{if } t_1 \leq t < t_2 \\ \sin(\omega_3 t), & \text{if } t_2 \leq t < t_3 \\ 0, & \text{otherwise} \end{cases} \tag{4.62}$$

where $\omega_1 < \omega_2 < \omega_3$. Also comment on the resolution of the various components present in the signal for various window lengths $L_1$, $L_2$, and $L_3$ by performing simulation. Consider the values of $t_1$, $t_2$, $t_3$, $\omega_1$, $\omega_2$, $\omega_3$, and suitable sampling frequency for signal $x(t)$ as per your choice for simulation.

Q 4.9 Compute the narrowband and wideband spectrograms for a linear chirp signal $x(t) = A\cos(2\pi(f_0 + \frac{1}{2}f_1 t)t)$. Consider the suitable values of the signal parameters and sampling rate for simulation. Also comment on the differences observed in these two obtained spectrograms.

Q 4.10 Load a speech signal from MATLAB using command load mtlb. Using this speech signal, compute the spectrogram using Hamming window of durations 5 ms and 25 ms with 50% overlap and comment on the differences in both of the spectrograms.

Q 4.11 Consider a quadratic chirp signal $x(t) = A\cos\left[2\pi(f_0 + \beta t^2)t\right]$. Perform the STFT analysis of this signal using a non-overlapping and overlapping window (with 50%, 70%, and 99% overlap). Also comment on the differences observed in the spectrogram of all the cases. Choose the parameters of signal $x(t)$ and sampling rate based on the suitability for analysis.

Q 4.12 Consider the two signals $x_1(t)$ and $x_2(t)$, $\forall t \in [0, 1]$ defined as following:

$$x_1(t) = \begin{cases} \cos(6\pi t)\cos(700\pi t), & \text{if } 0 \leq t < 0.5 \\ \cos(4\pi t)\cos(450\pi t), & \text{if } 0.5 \leq t \leq 1 \end{cases}$$

$$x_2(t) = \cos(6\pi t)\cos(700\pi t) + \cos(4\pi t)\cos(450\pi t)$$

Compute the spectrogram and sonogram for both signals. Choose the suitable sampling rate for both the signals.

# 5 Wavelet Transform

*"You have the right to work. But never to the fruit of work."* –The Bhagavad Gita

## 5.1 CONTINUOUS WAVELET TRANSFORM

The STFT provides TFA with the fixed resolution for all frequency components. The size of the time-frequency tiles or Heisenberg boxes in time-frequency plane is uniform. Due to this uniform resolution of the basis functions corresponding to all frequency components, there is an uniform resolution for all the frequency components in case of STFT. But, low-frequency components require good frequency resolution or poor time resolution, whereas high-frequency components require good time resolution and poor frequency resolution. This multi-resolution requirement motivates the development of continuous wavelet transform (CWT), which provides a multi-resolution analysis (MRA) for the signals [54, 48, 55]. The basis functions in CWT are designed from the mother wavelet denoted by $\psi(t)$. All other basis functions are obtained from the scaled and shifted versions of this mother wavelet. The mother wavelet $\psi(t)$ should satisfy following properties:

Property 1: The wavelet should have zero average value. This property of wavelet ensures the oscillatory nature. The mathematical expression for this property is given as,

$$\int_{-\infty}^{\infty} \psi(t)dt = 0$$

Property 2: The wavelet should be square integrable. Due to this property, the mother wavelet $\psi(t)$ has finite energy so that most of the energy is associated with a finite duration of $\psi(t)$. Mathematically, this property is expressed as,

$$\int_{-\infty}^{\infty} |\psi(t)|^2 dt = \text{Finite or } < \infty$$

Property 3: The wavelet should satisfy the admissibility condition, which helps in obtaining the inverse CWT or perfect reconstruction. Mathematically, this condition is expressed as,

$$\int_{-\infty}^{\infty} \frac{|\Psi(\omega)|^2}{|\omega|} d\omega = \text{Finite or } < \infty$$

where $\Psi(\omega)$ denotes the Fourier transform of the mother wavelet $\psi(t)$. Figure 5.1. shows plot of four mother wavelets namely, Haar, Symlet 4, Daubechies 3, and Mexican hat.

DOI: 10.1201/9781003367987-5

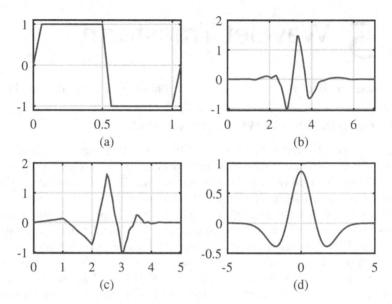

**Figure 5.1**   Plot of mother wavelets: (a) Haar, (b) Symlet 4 (c) Daubechies 3, and (d) Mexican hat.

The CWT of a continuous-time signal $s(t)$ is given as follows:

$$\mathrm{WT}(a,b) = \int_{-\infty}^{\infty} s(t) \frac{1}{\sqrt{a}} \psi^* \left( \frac{t-b}{a} \right) dt$$

$$= \int_{-\infty}^{\infty} s(t) \psi_{a,b}^*(t) dt \qquad (5.1)$$

Here, $\psi_{a,b}(t) = \frac{1}{\sqrt{a}} \psi \left( \frac{t-b}{a} \right)$ denotes the basis functions for CWT, parameter $b$ is known as shifting parameter or translation parameter, and $a$ represents a scaling or dilation parameter. In analysis expression (Eq. (5.1)), the normalization factor $\frac{1}{\sqrt{a}}$ in the basis functions makes same energy of all these basis functions for parameters $a$ and $b$. This means,

$$E = \int_{-\infty}^{\infty} \left| \psi_{a,b}(t) \right|^2 dt = \int_{-\infty}^{\infty} \left| \frac{1}{\sqrt{a}} \psi \left( \frac{t-b}{a} \right) \right|^2 dt$$

putting, $\frac{t-b}{a} = \alpha$; $dt = a d\alpha$ and after further simplification, we have,

$$E = \int_{-\infty}^{\infty} |\psi(t)|^2 dt$$

This means that energy of the mother wavelet $\psi(t)$ is same as the energy of the dilated and shifted wavelets $\psi_{a,b}(t)$ and this is due to the normalization factor $\frac{1}{\sqrt{a}}$. The dilated and shifted wavelets $\psi_{a,b}(t)$ are also known as daughter wavelets.

## 5.2    SCALOGRAM

Scalogram is obtained from the squared magnitude of CWT which is defined as,

$$\text{Scalogram} = |\text{WT}(a,b)|^2 = \left| \int_{-\infty}^{\infty} x(t) \frac{1}{\sqrt{a}} \psi^* \left( \frac{t-b}{a} \right) dt \right|^2 \qquad (5.2)$$

The scalogram provides time-scale representation of the signal. The scale can be converted into frequency and like spectrogram, we can obtain TFR using scalogram. Since scalogram represents the energy of the signal in time-scale or time-frequency domain, it can be plotted in the form of an image or contour plot. The scalogram of a pulse signal obtained using 'Morse' wavelet is shown in the Fig. 5.2. For simulation, sampling rate of 100 Hz has been considered. We can also explain MRA property from Fig. 5.2. It can be noticed from the scalogram of a pulse signal, which has all frequency components during transition. The high-frequency components are analyzed with good time-resolution (low scale) as compared to the low-frequency components (high scale).

## 5.3    FEATURES OF CWT

### 5.3.1    TIME-FREQUENCY LOCALIZATION

Suppose mother wavelet $\psi(t)$ has the following Heisenberg box:

$$[(m_t, m_\omega); (2\sigma_t, 2\sigma_\omega)]$$

Then, Heisenberg box for daughter wavelets, $\psi_{a,b}(t) = \frac{1}{\sqrt{|a|}} \psi \left( \frac{t-b}{a} \right)$ can be computed as (see Table 3.1),

$$\left[ \left( am_t + b, \frac{m_\omega}{a} \right) ; \left( 2a\sigma_t, \frac{2\sigma_\omega}{a} \right) \right]$$

From Hesenberg box for mother wavelet and daughter wavelet, it can be stated that the area of the Heisenberg box corresponding to mother wavelet is $4\sigma_t \sigma_\omega$ [51, 40]. The area of the Heisenberg box corresponding to daughter wavelets can be obtained as $4\sigma_t \sigma_\omega$ from the above-mentioned Heisenberg box.

The area of the Heisenberg box for mother and daughter wavelets is same. It should be noted that when the value of scale parameter is low then duration of the box in time-frequency plane is less. On the other hand, for low value of scale parameter $a$, the bandwidth of the Heisenberg box is higher. It indicates good time resolution and poor frequency resolution. On the opposite side, for high value of parameter $a$, duration of the Heisenberg box increases, so it provides poor time resolution and at the same time bandwidth of the Heisenberg box is reduced for high value of parameter $a$; so that it provides poor time resolution and good frequency resolution.

Figure 5.3 shows Heisenberg boxes for two sets of values of scale and shifting parameters $(a_1, b_1)$, $(a_2, b_2)$ for the daughter wavelets where $a_1 > a_2$ and $b_1 > b_2$.

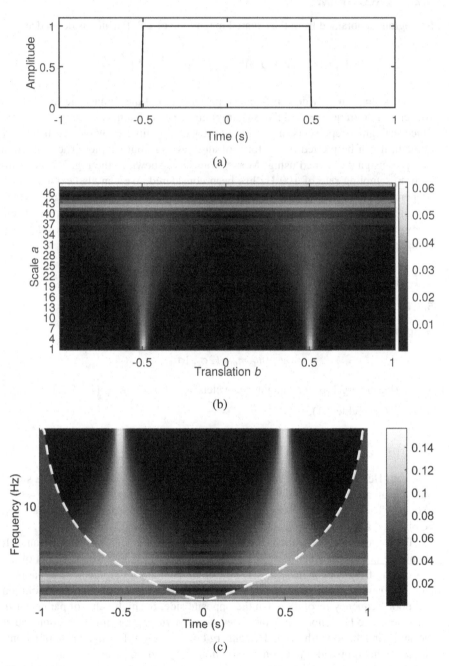

**Figure 5.2** (a) Pulse signal, (b) time-scale representation of pulse signal, and (c) TFR of pulse signal.

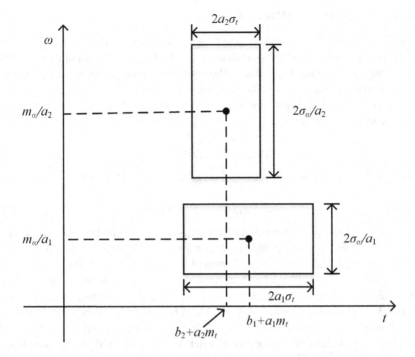

**Figure 5.3** Heisenberg boxes for two sets of scale and shifting parameters.

The time spread or duration of the daughter wavelet is proportional to the scale parameter ($a$) and the frequency spread or bandwidth of the daughter wavelet is proportional to the inverse of the scale parameter ($a$). Mean frequency is inversely proportional to scale parameter $a$, which can be expressed as,

$$\text{Frequency} \propto \frac{1}{\text{scale}}$$

For example, Heisenberg box for three scales $a = 1$, $a = 4$, and $a = \frac{1}{4}$, can be given as follows:

$$a = 1, \quad \text{Heisenberg box} = [(m_t + b, m_\omega); (2\sigma_t, 2\sigma_\omega)]$$

$$a = 4, \quad \text{Heisenberg box} = [(4m_t + b, \frac{m_\omega}{4}); (8\sigma_t, \frac{\sigma_\omega}{2})]$$

$$a = \frac{1}{4}, \quad \text{Heisenberg box} = [(\frac{m_t}{4} + b, 4m_\omega); (\frac{\sigma_t}{2}, 8\sigma_\omega)]$$

If $a = 1$ is considered as a reference then $a = 4$ provides relatively poor time resolution and good frequency resolution; and $a = \frac{1}{4}$ gives good time resolution and poor frequency resolution.

## 5.3.2   CONSTANT-Q ANALYSIS

The STFT provides constant-bandwidth analysis and due to that reason STFT-based analysis has fixed frequency resolution for all the frequency components present in the signal. On the other hand, due to the multi-resolution property, the CWT can be considered as constant-Q or octave analysis technique [56].

Suppose $\Psi(\omega)$ represents the Fourier transform of $\psi(t)$, then the Fourier transform of $\psi_{a,b}(t)$ is given by,

$$\mathscr{F}\{\psi_{a,b}(t)\} = \sqrt{a}e^{-j\omega b}\Psi(a\omega) \tag{5.3}$$

The squared magnitude of spectrum of daughter wavelets is given by,

$$|\mathscr{F}\{\psi_{a,b}(t)\}|^2 = a|\Psi(a\omega)|^2 \tag{5.4}$$

From Eq. (5.4), it can be seen that if there is an increase in the scale then there is a less spread of the spectrum of the daughter wavelet. On the other hand, decreasing the scale increases the spread of the spectrum of daughter wavelet. Such variable spread of spectrum or bandwidth of daughter wavelet helps in maintaining constant-Q analysis. Constant-Q property can also be understood with the frequency-domain localization parameters for daughter wavelet. Mean frequency and bandwidth of daughter wavelets are given by, $m'_\omega = \frac{m_\omega}{a}$ and $2\sigma'_\omega = \frac{2\sigma_\omega}{a}$, where $m_\omega$ and $2\sigma_\omega$ are the mean frequency and bandwidth of mother wavelet. Q-factor of the daughter wavelets can be expressed as $\frac{m_\omega}{2\sigma_\omega}$, which is constant.

Figures 5.4 (a) and (b) show the STFT basis functions and wavelet basis functions in time-domain and frequency-domain, respectively. It is clear from Fig. 5.4 (a) that the different frequency components have fixed resolution in time- and frequency-domain (fixed spread in time- and frequency-domain). On the other hand, Fig. 5.4 (b) shows the variable resolution in time- and frequency-domain (variable spreads in time- and frequency-domain) for different wavelet basis functions having different scale parameters.

## 5.3.3   CONVOLUTION FORM

The CWT can be represented in the form of convolution which is useful for filtering operation in wavelet-domain. Based on the Parseval's theorem, CWT of a signal $s(t)$ with Fourier transform $S(\omega)$, as given in Eq. (5.1), can be written as follows:

$$
\begin{aligned}
\text{WT}(a,b) &= \frac{1}{2\pi} \int_{-\infty}^{\infty} S(\omega) \left[ \sqrt{a}e^{-j\omega b}\Psi(a\omega) \right]^* d\omega \\
&= \frac{\sqrt{a}}{2\pi} \int_{\infty}^{\infty} [S(\omega)\Psi^*(a\omega)] e^{j\omega b} d\omega
\end{aligned}
\tag{5.5}
$$

This expression shows that the CWT can be computed using inverse Fourier transform of the band-limited spectrum. By replacing $b$ by $t$ and $t$ by $\alpha$, the Eq. (5.1) for CWT of signal $s(t)$ can be written as,

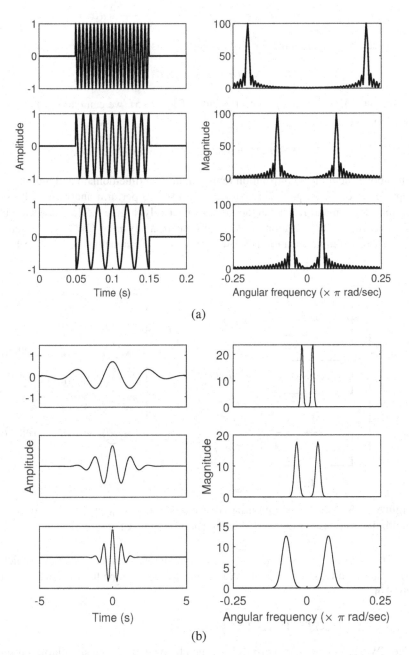

**Figure 5.4** (a) Three different basis functions of STFT and their corresponding spectrums, (b) wavelet basis functions for three scales and corresponding spectrums.

$$\text{WT}(a,t) = \int_{-\infty}^{\infty} s(\alpha) \frac{1}{\sqrt{a}} \psi^* \left( \frac{\alpha - t}{a} \right) d\alpha$$

which can be written in the form of convolution as,

$$\text{WT}(a,t) = s(t) * \psi_a^*(-t) \qquad (5.6)$$

where, $\psi_a^*(t) = \frac{1}{\sqrt{a}} \psi^* \left( \frac{t}{a} \right)$. With the help of Eq. (5.5), we can express Eq. (5.6) in frequency-domain as,

$$\mathscr{F}\{\text{WT}(a,t)\} = S(\omega)\Psi_a^*(\omega) = S(\omega)\sqrt{a}\Psi^*(a\omega) \qquad (5.7)$$

The above-mentioned convolution representation in time-domain and multiplication representation in frequency-domain can be used to perform analysis of the signal in time-domain and frequency-domain, respectively. Figure 5.5 (a) shows the block diagram for the convolution-based time-domain analysis of CWT. The frequency-domain-based multiplication is shown for analysis of signal using CWT in Fig. 5.5 (b).

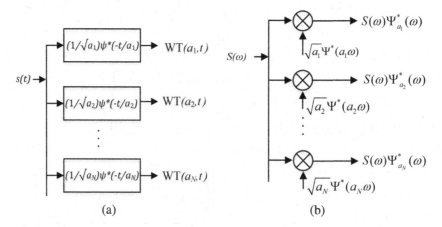

(a)                                    (b)

**Figure 5.5** (a) Convolution-based time-domain analysis of CWT and (b) multiplication-based frequency-domain analysis of CWT.

The analysis process of CWT in time-domain and frequency-domain, which has been shown in Figs. 5.5 (a) and (b), is useful for performing filtering in the wavelet-domain.

### 5.3.4 SINGULARITY DETECTION

The CWT has an interesting property which helps in detecting singularity present in the data when scale $a$ tends to be zero.

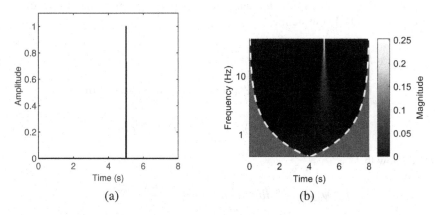

**Figure 5.6** (a) Impulse signal and (b) its TFR based on scalogram.

Suppose, for example, signal $x(t) = p\delta(t - T)$, then its CWT is given by Eq. (5.1),

$$\text{WT}(a,b) = \int_{-\infty}^{\infty} p\delta(t-T)\frac{1}{\sqrt{a}}\psi\left(\frac{t-b}{a}\right)dt$$

$$= \frac{p}{\sqrt{a}}\psi\left(\frac{T-b}{a}\right)$$

In case of an impulse function which is located at $t = T$, the corresponding CWT is also a wavelet scaled by $a$ and centered at $b$. When $a$ tends to zero, then CWT becomes narrow exactly at singularity and grows as $\frac{p}{\sqrt{a}}$, where $p$ is the strength of the impulse function.

For an example of singularity detection, CWT (using 'Morse' wavelet) of an impulse function, $x(t) = \delta(t-5)$ together with its scalogram have been shown in Fig. 5.6. The sampling rate of 100 samples/s and $\delta(t)$ as Kronecker delta are considered for simulation. It is clear from the figure that when scale becomes very less it shows narrow CWT in order to detect impulse located at $t = 5$ s.

## 5.4 INVERSE CWT

The inverse CWT can be performed by the following mathematical expression:

$$s(t) = \frac{1}{k}\int_0^{\infty}\int_{-\infty}^{\infty}\text{WT}(a,b)\psi_{a,b}(t)\frac{da\,db}{a^2} \tag{5.8}$$

Where parameter $k$ is a constant. The inverse CWT formula Eq. (5.8) can be proved as follows:

$$\text{WT}(a,b) = \frac{1}{2\pi}\int_{-\infty}^{\infty}S(\omega)\Psi^*_{a,b}(\omega)d\omega$$

$$= \frac{\sqrt{a}}{2\pi}\int_{-\infty}^{\infty}\Psi^*(a\omega)S(\omega)e^{jb\omega}d\omega$$

From Eq. (5.7), the CWT $WT(a,b)$, can be considered as inverse Fourier transform of $\sqrt{a}S(\omega)\Psi^*(a\omega)$. Let us define,

$$K(a) = \int_{-\infty}^{\infty} WT(a,b)\psi_{a,b}(t)db$$

$$K(a) = \frac{\sqrt{a}}{2\pi} \int_{-\infty}^{\infty} \left( \int_{-\infty}^{\infty} \Psi^*(a\omega)S(\omega)e^{jb\omega}d\omega \right) \psi_{a,b}(t)db$$

$$= \frac{\sqrt{a}}{2\pi} \int_{-\infty}^{\infty} \Psi^*(a\omega)S(\omega) \int_{-\infty}^{\infty} \psi_{a,b}(t)e^{jb\omega}db\,d\omega \qquad (5.9)$$

We know that,

$$\int_{-\infty}^{\infty} \psi_{a,b}(t)e^{jb\omega}db = \frac{1}{\sqrt{a}} \int_{-\infty}^{\infty} \psi\left(\frac{t-b}{a}\right) e^{jb\omega}db$$

Putting $b_1 = \frac{t-b}{a}$; $db_1 = \frac{-db}{a}$, and further simplification provides,

$$\int_{-\infty}^{\infty} \psi_{a,b}(t)e^{jb\omega}db = \sqrt{a}e^{j\omega t}\Psi(a\omega)$$

Now, from Eq. (5.9), we have,

$$K(a) = \frac{\sqrt{a}}{2\pi} \int_{-\infty}^{\infty} \Psi^*(a\omega)S(\omega)\sqrt{a}e^{j\omega t}\Psi(a\omega)d\omega$$

$$= \frac{a}{2\pi} \int_{-\infty}^{\infty} |\Psi(a\omega)|^2 S(\omega)e^{j\omega t}d\omega$$

Now, $$\int_{-\infty}^{\infty} K(a)\frac{da}{a^2} = \frac{1}{2\pi} \int_{-\infty}^{\infty} \int_{-\infty}^{\infty} \frac{|\Psi(a\omega)|^2}{a} S(\omega)e^{j\omega t}d\omega\,da$$

and $$\int_{-\infty}^{\infty} \frac{|\Psi(a\omega)|^2}{a}da = k$$

Therefore, $$\int_{-\infty}^{\infty} K(a)\frac{da}{a^2} = \frac{1}{2\pi} \int_{-\infty}^{\infty} kS(\omega)e^{j\omega t}d\omega$$

$$= ks(t)$$

We have, $s(t) = \frac{1}{k} \int_{-\infty}^{\infty} K(a)\frac{da}{a^2}$. The constant parameter $k$ needs to be finite in order to have reconstruction of the signal from CWT.

## 5.5   SOME PROPERTIES OF CWT

The CWT has some interesting properties which are useful for analysis of a signal and these properties are listed in Table 5.1.

## 5.6   ENERGY CONSERVATION IN CWT

If signal $s(t)$ has CWT, $WT(a,b)$, then according to this principle, energy in time-domain equals to the energy in wavelet-domain. Mathematically, it can be expressed

**Table 5.1**
**Properties of CWT**

| Property | Mathematical Expression |
|---|---|
| Linearity | If $x(t) \leftrightarrow \mathrm{WT}_x(a,b)$ then, $$\sum_{k=1}^{N} c_k x_k(t) \leftrightarrow \sum_{k=1}^{N} c_k \mathrm{WT}_{x_k}(a,b)$$ Here, $c_1, c_2, \ldots, c_N$ are constants. |
| Time shifting | If $x_1(t) = x(t - \tau)$ then, $\mathrm{WT}_{x_1}(a,b) = \mathrm{WT}_x(a, b - \tau)$ |
| Scaling | If $x_1(t) = \frac{1}{\sqrt{a_1}} x\left(\frac{t}{a_1}\right)$ then, $\mathrm{WT}_{x_1} = \mathrm{WT}_x\left(\frac{a}{a_1}, \frac{b}{a_1}\right)$ |
| Wavelet shifting | If $\psi_1(t) = \psi(t - \tau)$ then, $\mathrm{WT}_1(a,b) = \mathrm{WT}(a, b + a\tau)$ |
| Wavelet scaling | If $\psi_1(t) = \frac{1}{\sqrt{a_1}} \psi\left(\frac{t}{a_1}\right)$ then, $\mathrm{WT}_1(a,b) = \mathrm{WT}(aa_1, b)$ |
| Linear combination of wavelets | If $\psi_T = \sum_{k=1}^{N} c_k \psi_k(t)$ then, $$\mathrm{WT}_T(a,b) = \sum_{k=1}^{N} c_k \mathrm{WT}_k(a,b)$$ |

as,

$$\int_{-\infty}^{\infty} |s(t)|^2 dt = \frac{1}{k} \int_{0}^{\infty} \int_{-\infty}^{\infty} |\mathrm{WT}(a,b)|^2 \frac{da\,db}{a^2} \tag{5.10}$$

The above-mentioned energy conservation rule for CWT in Eq. (5.10) can be proved as, taking RHS of Eq. (5.10),

$$\mathrm{RHS} = \frac{1}{k} \int_{0}^{\infty} \int_{-\infty}^{\infty} |\mathrm{WT}(a,b)|^2 \frac{da\,db}{a^2}$$

Using the Parseval's theorem, we can write RHS as follows:

$$\mathrm{RHS} = \frac{1}{k} \int_{0}^{\infty} \frac{1}{2\pi} \int_{-\infty}^{\infty} |\sqrt{a} S(\omega) \Psi^*(a\omega)|^2 \frac{da\,d\omega}{a^2}$$

$$= \frac{1}{k} \int_{0}^{\infty} \frac{1}{2\pi} \int_{-\infty}^{\infty} a |S(\omega)|^2 |\Psi(a\omega)|^2 d\omega \frac{da}{a^2}$$

$$= \frac{1}{k} \int_{-\infty}^{\infty} \frac{1}{2\pi} |S(\omega)|^2 \int_{0}^{\infty} \frac{|\Psi(a\omega)|^2 da}{a} d\omega$$

$$= \frac{1}{2\pi} \int_{-\infty}^{\infty} |S(\omega)|^2 d\omega \quad = \int_{-\infty}^{\infty} |s(t)|^2 dt$$

Hence, the energy of the signal in time-domain equals to the energy in frequency-domain and CWT domain, which can be considered as Parseval's theorem for CWT.

## 5.7  WAVELET SERIES

### 5.7.1  CONCEPT OF FRAMES

In CWT, the shifting and scale parameters are continuous in nature and require infinite memory in order to store them in computer. Due to these reasons, discretization of these parameters is required. This discretization can be considered as optimal when the representation requires less number of wavelet basis functions. But in signal analysis and processing, more number of basis functions are required than the absolutely necessary requirement. The requirement of such basis functions leads to the concept of frames.

It should be noted that the reconstruction formula in CWT requires double integration over $b$ and $a$ (shifting and scale parameters) and requires over complete set of basis functions which is highly redundant. Therefore, the discretization of these parameters is used as follows:

$$a = a_0^m, \quad a_0 \neq 1 \text{ and } m \in \mathbb{Z}$$
$$b = nb_0 a_0^m, \quad b_0 \neq 0 \text{ and } m, n \in \mathbb{Z}$$

where $a_0$ is a constant and $b_0$ is a positive constant.

After this discretization, the daughter wavelets $\psi_{a,b}(t)$ are mathematically represented as,

$$\psi_{m,n}(t) = a_0^{\frac{-m}{2}} \psi(a_0^{-m} t - nb_0) \tag{5.11}$$

Generally, the dyadic sampling ($a_0 = 2$ and $b_0 = 1$) is used for discretization of the shifting and scale parameters of daughter wavelets. For dyadic sampling, daughter wavelets are expressed as,

$$\psi_{m,n}(t) = 2^{\frac{-m}{2}} \psi(2^{-m} t - n) \tag{5.12}$$

The concept of frames provides a set of conditions which helps in obtaining a stable reconstruction. The reconstruction of signal $s(t)$ from the wavelets corresponding to these discretetized set of parameters is given by following mathematical expression:

$$s(t) = \sum_{m=-\infty}^{\infty} \sum_{n=-\infty}^{\infty} \text{WT}(m,n) \psi_{m,n}(t) \tag{5.13}$$

Then a family of functions $\psi_{m,n}(t)$ which belongs to Hilbert space is known as a frame if there exist parameters $C$ and $D$ with the condition, $0 < C \leq D$, so that for the signal $s(t)$, we have the following mathematical relations:

$$C||s(t)||_2^2 \leq \sum_m \sum_n |\text{WT}(m,n)|^2 \leq D||s(t)||_2^2$$

Where $||s(t)||_2^2 = \int_{-\infty}^{\infty} |s(t)|^2 dt$ and $\quad WT(m,n) = \langle s(t), \psi_{m,n}(t) \rangle$; the parameters $C$ and $D$ are called frame bounds.

If $C = D$, then this case is known as tight frame condition. The redundancy ratio is defined as the ratio of these parameters as, $C/D$; which is equal to 1 for tight frame condition. Together with tight frame condition, if the basis functions are normalized so that $||\psi_{m,n}(t)||_2^2 = 1$ and $C = D = 1$ then the set of $\psi_{m,n}(t)$ consists of orthonormal basis functions. It should be noted that the $||s(t)||_2^2$ is the total energy of the signal $s(t)$ in time-domain and $|WT(m,n)|^2$ is also related to the energy in wavelet-domain and in tight frame condition, the energy of the signal in time-domain equals to the energy in the transform-domain, i.e., wavelet-domain.

The physical meaning of the condition, $\sum_m \sum_n |WT(m,n)|^2 \le D||s(t)||_2^2$ is that $\sum_m \sum_n |WT(m,n)|^2$ should be finite or less than $\infty$ in order to have proper representation of the signal. Similarly, the condition, $C||s(t)||_2^2 \le \sum_m \sum_n |WT(m,n)|^2$, makes sure that energy of the signal cannot be zero; which means that $C$ is a positive parameter.

### 5.7.2 DYADIC SAMPLING

The daughter wavelets with discretized parameters, $\psi_{m,n}(t) = a_0^{\frac{-m}{2}} \psi(a_0^{-m} t - nb_0)$, constitute a frame in Hilbert space, then the frame bounds conditions can be expressed as [12, 57],

$$\frac{b_0 \log(a_0)}{2\pi} C \le \int_0^\infty \frac{|\Psi(\omega)|^2}{\omega} d\omega \le \frac{b_0 \log(a_0)}{2\pi} D \qquad (5.14)$$

This condition is based on the admissibility condition of the mother wavelet. When $\{\psi_{m,n}(t)\}$ is a tight frame, the Eq. (5.14) can be written as,

$$C = \frac{2\pi}{b_0 \log(a_0)} \int_0^\infty \frac{|\Psi(\omega)|^2}{\omega} d\omega = D \qquad (5.15)$$

In case of $\{\psi_{m,n}(t)\}$ forms a set of orthonormal basis, the Eq. (5.15) can be expressed as,

$$\frac{b_0 \log(a_0)}{2\pi} = \int_0^\infty \frac{|\Psi(\omega)|^2}{\omega} d\omega$$

In STFT, the time-frequency plane is represented with rectangular sampling grid of the form $(m\omega_0, n\tau_0)$. The sampling grid in time-frequency plane for STFT is shown in Fig. 5.7 (a). In the case of wavelet transform, dyadic sampling grid is used to represent the time-scale plane. The dyadic sampling grid-based time-scale representation is shown in Fig. 5.7 (b) for wavelet transform.

In this dyadic sampling grid, basis functions corresponding to large scale shift with larger intervals. On the other hand, basis functions of low scale shift in smaller intervals. In order to perform sampling of time-scale plane $(a,b)$ with high resolution, $a_0$ and $b_0$ should be close to 1 and 0, respectively.

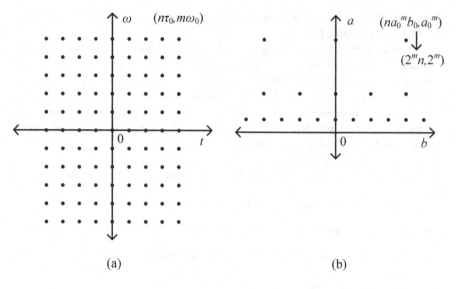

**Figure 5.7** Sampling grids for (a) STFT method and (b) dyadic wavelet transform.

### 5.7.3 WAVELET SERIES AS FILTER BANK

The orthonormal condition for wavelets which reduces the redundancy from the wavelet series can be represented as [12],

$$\langle \psi_{m,n}(t), \psi_{m_1,n_1}(t) \rangle = \delta(m - m_1)\delta(n - n_1), \quad \text{where,} \ m,n,m_1,n_1 \in \mathbb{Z} \qquad (5.16)$$

The orthonormal wavelets can be used as basis functions to represent the signal $s(t)$ as shown in Eq. (5.13). The wavelet coefficients $\mathrm{WT}(m,n)$ can be computed as inner product of the signal $s(t)$ and the wavelets $\psi_{m,n}(t)$ as,

$$\mathrm{WT}(m,n) = \langle s(t), \ \psi_{m,n}(t) \rangle$$

The computation of wavelet transform coefficients corresponding to dyadic sampling can be carried out by performing convolution of $s(t)$ and $\psi_m(t)$ as given in Eq. (5.6). Continuous wavelet coefficients $\mathrm{WT}(a,b)$ can be discretized by substituting $a = 2^m$ and $b = n2^m$ as,

$$\mathrm{WT}(m,n2^m) = 2^{-m/2} \int s(\tau)\psi^*(2^{-m}\tau - n)d\tau$$

The wavelet transform is implemented based on the filter bank, which is shown in Fig. 5.8.

It should be noted that the wavelet down-sampler changes with respect to scale in each branch of the filter bank. On the other hand, in STFT the same down-sampler

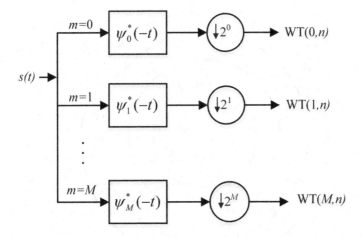

**Figure 5.8**   Wavelet series as filter bank.

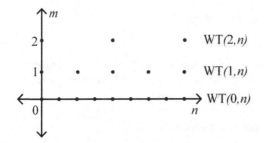

**Figure 5.9**   Dyadic wavelet series coefficients.

is used in each branch of the filter bank. The wavelet series coefficients are shown in Fig. 5.9.

It should be noted that as $m$ increases by one, the wavelet basis expands with a factor of two and when $n$ increases, the wavelet shifts toward right side. A specific daughter wavelet function represents a particular level of detail in the signal. When $m$ decreases, the wavelet basis will be compressed and the wavelet becomes fine grained which provides higher level of detail.

## 5.8   DISCRETE WAVELET TRANSFORM

### 5.8.1   MULTIRESOLUTION ANALYSIS

Wavelets work as band-pass filter, when scaling parameter increases the pass-band gets narrower and shifts toward lower frequency. As a result, ideally infinite number of wavelets will be required to cover the whole range of frequency band. In such scenario, scaling function with low-pass characteristics can replace large number of wavelets with high scale. This approach has been used in MRA, which helps to find

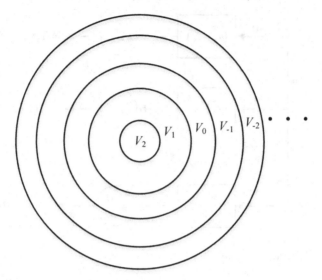

**Figure 5.10** Nesting of subspaces.

the signal components at different scales [9]. An infinite sequence $\{V_m\}$ corresponding to linear function spaces is termed as a MRA related to a scaling function $\phi(t)$ when following conditions are satisfied:

Condition 1: The spaces are nested as,

$$\underset{\text{coarse}}{\longleftarrow} \{0\} \subset \cdots \subset V_2 \subset V_1 \subset V_0 \subset V_{-1} \subset V_{-2} \subset \cdots \subset L^2(\mathscr{R}) \underset{\text{Fine}}{\longrightarrow}$$

Condition 2: $s(t) \in V_m$ if and only if,

$$s(2t) \in V_{m-1}$$

Condition 3: Functions $\phi_{0,n}(t) = \phi_n(t) = \phi(t-n)$ constitute an orthonormal basis for space $V_0$. The set, $\{\phi_{m,n}(t) = 2^{-m/2}\phi(2^{-m}t - n)\}; n \in \mathscr{Z}$ forms an orthonormal basis for $V_m$.

The above described spaces can also be sketched in terms of subspaces using onion type of layers in Fig. 5.10.

The best approximation at coarseness-level $m$ for signal $s(t)$, i.e., in $V_m$ can be given by following expression:

$$s_m(t) = \sum_{n=-\infty}^{\infty} s_{m,n}\phi_{m,n}(t) \tag{5.17}$$

$$\text{where, } s_{m,n} = \,< s(t), \phi_{m,n}(t) > \tag{5.18}$$

For example, Haar scaling function-based approximation of the signal $s(t)$ is obtained as follows:

Haar scaling function is defined as,

$$\phi(t) = \begin{cases} 1, & 0 \leq t < 1 \\ 0, & \text{otherwise} \end{cases} \tag{5.19}$$

and $\phi_{m,n}(t)$ can be expressed as,

$$\phi_{m,n}(t) = \begin{cases} 2^{-m/2}, & 2^m n \leq t < 2^m(n+1) \\ 0, & \text{otherwise} \end{cases} \tag{5.20}$$

The Haar scaling function and corresponding family are shown in Figs. 5.11 (a) and (b).

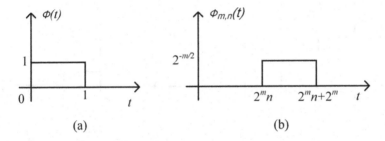

**Figure 5.11**   (a) Haar scaling function and (b) corresponding family.

The approximation of the signal $s(t)$ in $V_m$ can be computed using Eq. (5.18) as,

$$s_{m,n} = \int_{n2^m}^{(n+1)2^m} 2^{\frac{-m}{2}} s(t) dt$$

The approximation signal $s_m(t)$ can be computed from these coefficients using Eq. (5.17), as follows:

$$s_m(t) = \sum_{n=-\infty}^{\infty} \int_{n2^m}^{(n+1)2^m} 2^{-m/2} s(t) dt \, \phi_{m,n}(t)$$

$$= \sum_{n=-\infty}^{\infty} \underbrace{\frac{1}{2^m} \int_{n2^m}^{(n+1)2^m} s(t) dt}_{\text{Part 1}} \underbrace{2^{\frac{m}{2}} \phi_{m,n}(t)}_{\text{Part 2}}$$

In the expression above, Part 1 represents the average value of $s(t)$ for the interval $n2^m$ to $(n+1)2^m$ and Part 2 is a scaling function of height 1 for all $m$. It is clear that the approximation signal is obtained from the average value of the signal in each interval of width $2^m$ and approximating this function by a constant function (one) over that interval. Figure 5.12 shows one example where one arbitrary signal belongs to $V_0$ space, has been approximated in two different spaces $V_1$ and $V_2$. The scaling function corresponding to these spaces are shown in the right side of the figure. For clear visualization, the intervals for obtaining average value of the signal corresponding to two subspaces have been shown using dashed line.

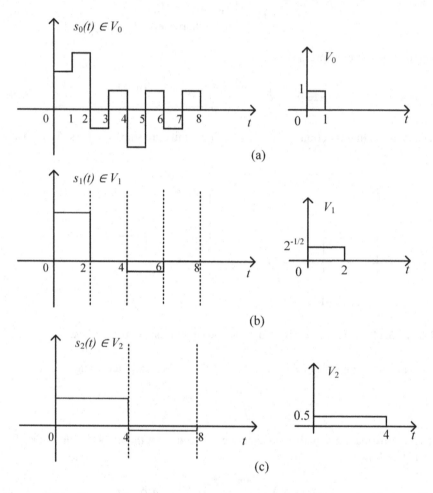

**Figure 5.12** Haar scaling functions based approximated signals (a) $s_0(t)$, (b) $s_1(t)$, and (c) $s_2(t)$, which are part of $V_0, V_1$, and $V_2$ subspaces, respectively.

### 5.8.2 TWO-SCALE RELATION FOR SCALING AND WAVELET FUNCTIONS

As $V_m \subset V_{m-1}$, which means that $V_{m-1}$ is more detailed as compared to $V_m$. Hence, in order to represent $\phi_{m,0}(t) \in V_{m-1}$ in space $V_m$, there should be existing set of coefficients $h(n)$ so that $\phi_{m,0}(t)$ can be expressed in terms of $\phi_{m-1,n}(t)$ as

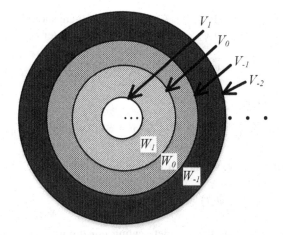

**Figure 5.13**  Detail subspaces and orthonormal components.

follows [58]:

$$\phi_{m,0}(t) = \sum_{n=-\infty}^{\infty} h(n)\phi_{m-1,n}(t)$$

It can be written as,

$$2^{-m/2}\phi\left(2^{-m}t\right) = \sum_{n=-\infty}^{\infty} h(n)2^{-(m-1)/2}\phi(2^{-(m-1)}t - n)$$

By putting $2^{-m}t = t$, the above expression can be written as,

$$\phi(t) = \sqrt{2} \sum_{n=-\infty}^{\infty} h(n)\phi(2t - n) \tag{5.21}$$

This equation is termed as scaling equation or two-scale relation for scaling function. This equation states that scaling function $\phi(t)$ can be represented as a linear combination of scaling functions itself, translated and scaled by a factor of 2.

$W_m$ represents the difference between subspaces $V_{m-1}$ and $V_m$, which is orthogonal component of $V_m$ in $V_{m-1}$. Figure 5.13 shows the different detail subspaces and associated orthogonal components.

It should be noted that,

$$V_{m-1} = V_m \oplus W_m, W_m = V_m^{\perp}$$

where $\oplus$ refers to the direct sum operation

$$V_m \cap W_m = \{0\}$$

and $\quad L^2(\mathbb{R}) = \cdots \oplus W_2 \oplus W_1 \oplus W_0 \oplus W_{-1} \oplus W_{-2} \oplus \cdots$

The signal $s(t)$ can be represented in terms of orthogonal components, $s_m(t)$ belong to space $W_m$ as,

$$s(t) = \cdots + s_1(t) + s_0(t) + s_{-1}(t) + \cdots$$

where, $s_m(t) = \sum_n \text{WT}(m,n)\psi_{m,n}(t)$. As $W_m \subset V_{m-1}$, there should be a set of coefficients $\{g(n)\}$; $n \in \mathbb{Z}$, so that $\psi_{m,0}(t)$ can be represented in terms of $\phi_{m-1,n}(t)$.

$$\psi_{m,0}(t) = \sum_{n=-\infty}^{\infty} g(n)\phi_{m-1,n}(t)$$

which can be written as,

$$2^{-m/2}\psi\left(2^{-m}t\right) = \sum_{n=-\infty}^{\infty} g(n)2^{-(m-1)/2}\phi\left(2^{-(m-1)}t - n\right)$$

By substituting $\frac{t}{2^m} = t$, we have following:

$$\psi(t) = \sqrt{2} \sum_{n=-\infty}^{\infty} g(n)\phi(2t - n) \tag{5.22}$$

This equation is termed as the wavelet scaling equation or two-scale relation for wavelet. To be valid scaling function and wavelet, Eqs. (5.21) and (5.22) have to be satisfied for some set of $h(n)$ and $g(n)$, respectively.

### 5.8.3  CONDITIONS ON $h(n)$ AND $g(n)$

The condition for $h(n)$ is given by following expression [55, 59]:

$$\delta(k) = \sum_{n=-\infty}^{\infty} h(n)h(n - 2k) \tag{5.23}$$

This condition can be proved as,
Consider LHS of Eq. (5.23), $\delta(k)$ which can be written as,

$$\delta(k) = \int_{-\infty}^{\infty} \phi(t)\phi(t - k)dt$$

$$= 2\sum_n h(n)\sum_l h(l) \int \phi(2t - n)\phi(2t - 2k - l)dt$$

$$= \sum_n h(n)\sum_l h(l) < \phi_{-1,n}(t), \phi_{-1,2k+l}(t) >$$

$$= \sum_n h(n)\sum_l h(l)\delta(n - 2k - l)$$

$$= \sum_n h(n)h(n - 2k)$$

Another condition to be met by coefficients $h(n)$ is $\sum_{n=-\infty}^{\infty} h(n) = \sqrt{2}$. This condition can be proved by performing integration on two-scale relation as follows:

$$\int_{-\infty}^{\infty} \phi(t)dt = \int_{-\infty}^{\infty} \sqrt{2} \sum_{n=-\infty}^{\infty} h(n)\phi(2t - n)dt$$

$$= \sqrt{2} \sum_{n=-\infty}^{\infty} h(n) \int_{-\infty}^{\infty} \phi(2t - n)dt$$

By putting, $2t - n = p$, and for any specific value of $n$ we have,

$$\int_{-\infty}^{\infty} \phi(t)dt = 2^{-1/2} \sum_{n=-\infty}^{\infty} h(n) \int_{-\infty}^{\infty} \phi(t)dt$$

The above expression holds true only when $\sum_{n=-\infty}^{\infty} h(n) = \sqrt{2}$.

The required condition for $g(n)$ can be expressed as,

$$\delta(k) = \sum_{n=-\infty}^{\infty} g(n)g(n-2k)$$

The above-mentioned condition can be proved using orthogonality between $V_m$ and $W_m$.

$$\delta(k) = \langle \psi_{m,0}(t), \phi_{m,k}(t) \rangle$$
$$= \left\langle \sum_n g(n)\phi_{m-1,n}(t), \sum_l h(l)\phi_{m-1,l+2k}(t) \right\rangle$$
$$= \sum_n g(n) \sum_l h(l) \langle \phi_{m-1,n}(t), \phi_{m-1,l+2k}(t) \rangle$$

Here, $\langle \phi_{m-1,n}(t), \phi_{m-1,l+2k}(t) \rangle = \delta(n - (l+2k))$
$$= \sum_n g(n)h(n-2k)$$

Another condition for $g(n)$ is $\sum_{n=-\infty}^{\infty} g(n) = 0$ which can be proved by performing integration on two-scale relation for wavelet as,

$$\int_{-\infty}^{\infty} \psi(t)dt = \int_{-\infty}^{\infty} \sqrt{2} \sum_{n=-\infty}^{\infty} g(n)\phi(2t-n)dt$$
$$= \sqrt{2} \sum_{n=-\infty}^{\infty} g(n) \int_{-\infty}^{\infty} \phi(2t-n)dt$$

By substituting, $2t - n = p$, and for any specific value of $n$ we have,

$$\int_{-\infty}^{\infty} \psi(t)dt = 2^{-1/2} \sum_{n=-\infty}^{\infty} g(n) \int_{-\infty}^{\infty} \phi(t)dt \qquad (5.24)$$

Scaling function $\phi(t)$ is a low-pass filter impulse response, so $\int_{-\infty}^{\infty} \phi(t)dt = 1$. It should be noted that $\int_{-\infty}^{\infty} \psi(t)dt = 0$. Hence, from Eq. (5.24), it can be seen $\sum_{n=-\infty}^{\infty} g(n) = 0$.

As an example, the Haar scaling function described in section 5.8.1, can be used for the computation of $g(n)$ and $h(n)$ for Haar wavelet and scaling functions as follows:

$$\langle \phi_{m-1,k}(t), \phi_{m,0}(t) \rangle = \left\langle \phi_{m-1,k}(t), \sum_n h(n)\phi_{m-1,n}(t) \right\rangle$$
$$= \sum_n h(n) \langle \phi_{m-1,k}(t), \phi_{m-1,n}(t) \rangle$$
$$= \sum_n h(n)\delta(n-k)$$
$$= h(k)$$

From the above relation $h(k)$ can be obtained for Haar scaling function, described in Eq. (5.20) as follows:

$$h(k) = \int_{-\infty}^{\infty} \phi_{m,0}(t)\phi_{m-1,k}(t)dt$$
$$= \int_{k2^{m-1}}^{(k+1)2^{m-1}} 2^{-(m-1)/2}\phi_{m,0}(t)dt$$

$h(k)$ for different values of $k$ can be obtained using the above relation and Eq. (5.20), which is given in following equation:

$$h(k) = \begin{cases} \frac{1}{\sqrt{2}}, & \text{for } k = 0,1 \\ 0, & \text{otherwise} \end{cases} \tag{5.25}$$

The relation between $g(n)$ and $h(n)$ can be expressed as [58, 60],

$$g(n) = (-1)^n h(N-n), \quad \text{for odd } N \tag{5.26}$$

Suppose select $N = 1$, then,

$$g(n) = (-1)^n h(N-n)$$
$$= (-1)^n h(1-n)$$

The coefficients $g(n)$ can be computed using the above equation, which is given as,

$$g(n) = \begin{cases} \frac{1}{\sqrt{2}}, & n = 0 \\ -\frac{1}{\sqrt{2}}, & n = 1 \\ 0, & \text{otherwise} \end{cases}$$

Now, $\psi(t)$ can be computed as,

$$\psi(t) = \sqrt{2} \sum_{n=-\infty}^{\infty} g(n)\phi(2t - n)$$

$$= \phi(2t) - \phi(2t - 1)$$

$$= \begin{cases} 1; & 0 \le t < \frac{1}{2} \\ -1; & \frac{1}{2} \le t < 1 \\ 0; & \text{otherwise} \end{cases}$$

The family of mother wavelet, $\psi_{m,n}(t)$ can be computed using the expression $\psi(t)$ as,

$$\psi_{m,n}(t) = \begin{cases} 2^{-m/2}; & n2^m \le t < (n+\frac{1}{2})2^m \\ -2^{m/2}; & (n+\frac{1}{2})2^m \le t < (n+1)2^m \\ 0; & \text{otherwise} \end{cases}$$

The plots of $\psi(t)$ and $\psi_{m,n}(t)$ are shown in Figs. 5.14 (a) and (b).

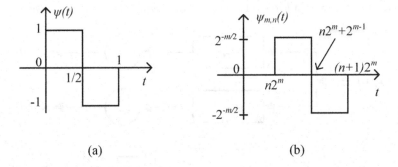

(a)                                    (b)

**Figure 5.14**   (a) Haar mother wavelet function and (b) corresponding family.

## 5.9   DWT BASED ON FILTER BANK

Suppose, the best approximation of the signal $s(t)$ is $s_0(t)$ corresponding $0^{\text{th}}$ level of coarseness. The signal $s_0(t)$ can be decomposed into scaling coefficients and wavelet coefficients at higher levels. As, $s_0(t) \in V_0$ and $V_0 = V_1 \oplus W_1$, the function $\phi_{0,n}(t)$ in space $V_0$ can be written as a linear combination of the basis functions which are associated with the spaces $V_1$ and $W_1$ [12, 59]. The corresponding coefficients are denoted as, $a_0(n)$, $a_1(n)$, and $d_1(n)$. The signal $s_0(t)$ can be expressed as follows:

$$s_0(t) = \sum_n a_0(n)\phi_{0,n}(t)$$

$$= \sum_n a_1(n)\phi_{1,n}(t) + \sum_n d_1(n)\psi_{1,n}(t)$$

Here,

$$a_1(n) = \langle s_0(t), \phi_{1,n}(t) \rangle$$

$$= \left\langle \sum_k a_0(k)\phi_{0,k}(t), \phi_{1,n}(t) \right\rangle$$

$$= \sum_k a_0(k) \langle \phi_{0,k}(t), \phi_{1,n}(t) \rangle$$

$$= \sum_k a_0(k) \left\langle \phi_{0,k}(t), \sum_l h(l)\phi_{0,l+2n}(t) \right\rangle$$

$$= \sum_k a_0(k) \sum_l h(l) \langle \phi_{0,k}(t), \phi_{0,l+2n}(t) \rangle,$$

where $\langle \phi_{0,k}(t), \phi_{0,l+2n}(t) \rangle = \delta(k - l - 2n)$

$$= \sum_k a_0(k) h(k - 2n)$$

Therefore, $a_1(n)$ can be considered as the output of the convolution of $a_0(n)$ with a time reversed version of $h(n)$ and after that down sampling by factor of two.

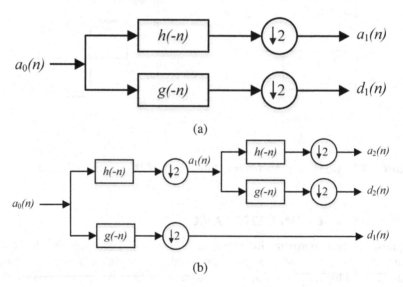

(a)

(b)

**Figure 5.15** (a) Detail and approximation coefficients computation for DWT for level-1 and (b) analysis filter bank for DWT for level-2.

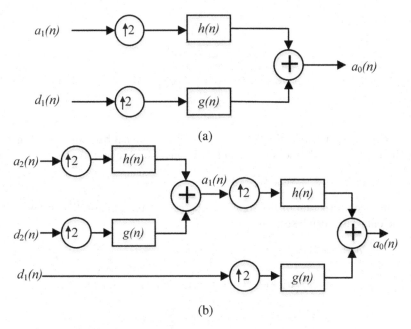

**Figure 5.16** (a) Reconstruction process for 1st level decomposition and (b) reconstruction process for 2nd level decomposition.

The $d_1(n)$ can be expressed as,

$$d_1(n) = \langle s_0(t), \psi_{1,n}(t) \rangle$$

$$= \left\langle \sum_k a_0(k)\phi_{0,k}(t), \psi_{1,n}(t) \right\rangle$$

$$= \sum_k a_0(k) \langle \phi_{0,k}(t), \psi_{1,n}(t) \rangle$$

$$= \sum_k a_0(k) \left\langle \phi_{0,k}(t), \sum_l g(l)\phi_{0,l+2n}(t) \right\rangle$$

$$= \sum_k a_0(k) \sum_l g(l) \langle \phi_{0,k}(t), \phi_{0,l+2n}(t) \rangle$$

where $\langle \phi_{0,k}(t), \phi_{0,l+2n}(t) \rangle = \delta(k - l - 2n)$

$$= \sum_k a_0(k) g(k - 2n)$$

Here, it should be noted that $d_1(n)$ is the obtained output using the convolution process of $a_0(n)$ with a time-reversed version of $g(n)$ and then performing down sampling by a factor of two. The above-mentioned processes together for obtaining $a_1(n)$ and $d_1(n)$ from the $a_0(n)$ provide a level of DWT which has been shown in Fig. 5.15 (a). The same process can be written in the generalized form for next higher level of coefficients as,

$$a_{m+1}(n) = \sum_k a_m(k)h(k-2n)$$

$$d_{m+1}(n) = \sum_k a_m(k)g(k-2n)$$

For example, analysis filter bank for two-level DWT has been shown in Fig. 5.15 (b). In the form of subspaces $m^{th}$ level DWT can be written as,

$$V_0 = W_1 \oplus W_2 \oplus W_3 \oplus \ldots \oplus W_m \oplus V_m$$

Now, the reconstruction of the signal can be carried out in the following way: The coefficients $a_0(n)$ can be expressed as,

$$
\begin{aligned}
a_0(n) &= \langle s_0(t), \phi_{0,n}(t) \rangle \\
&= \left\langle \sum_k a_1(k)\phi_{1,k}(t) + \sum_k d_1(k)\psi_{1,k}(t), \phi_{0,n}(t) \right\rangle \\
&= \sum_k a_1(k) \langle \phi_{1,k}(t), \phi_{0,n}(t) \rangle + \sum_k d_1(k) \langle \psi_{1,k}(t), \phi_{0,n}(t) \rangle \\
&= \sum_k a_1(k)h(n-2k) + \sum_k d_1(k)g(n-2k)
\end{aligned}
$$

This process can be implemented with the help of filter bank, and synthesis process for $1^{st}$ level decomposition is shown in Fig. 5.16 (a).

The above-mentioned process can be applied to obtain $a_1(n)$ from $a_2(n)$ and $d_2(n)$ as,

$$a_1(n) = \sum_k a_2(k)h(n-2k) + \sum_k d_2(k)g(n-2k)$$

The resulting reconstruction process has been sketched in Fig. 5.16 (b) for level-2 DWT. The process of reconstruction can be generalized for any level of DWT. It should be noted that,

$$
\begin{aligned}
a_0(n) &= \langle s(t), \phi_{0,n}(t) \rangle \\
&= \langle s(t), \phi(t-n) \rangle \approx s(n)
\end{aligned}
$$

If $s(t)$ is smooth signal and scaling function in space $V_0$, $\phi(t)$, is of sufficiently short duration and $\phi(t)$ has sufficiently fine resolution as compared to resolution of $s(t)$. Then the sampled signal is sufficient for DWT.

In case $V_0$ is not fine enough then the signal $s(t)$ can be approximated by considering sufficiently fine resolution space $V_m$ with $m$ is large and negative [56], so that,

$$\left\langle 2^{-\frac{m}{2}} \phi(2^{-m}t - n), s(t) \right\rangle \approx 2^{-\frac{m}{2}} s(2^m n)$$

Figure 5.17 shows 4-level DWT-based decomposition for an EEG signal obtained from motor movement/imagery dataset (Physionet ATM) [3, 7]. The EEG signal

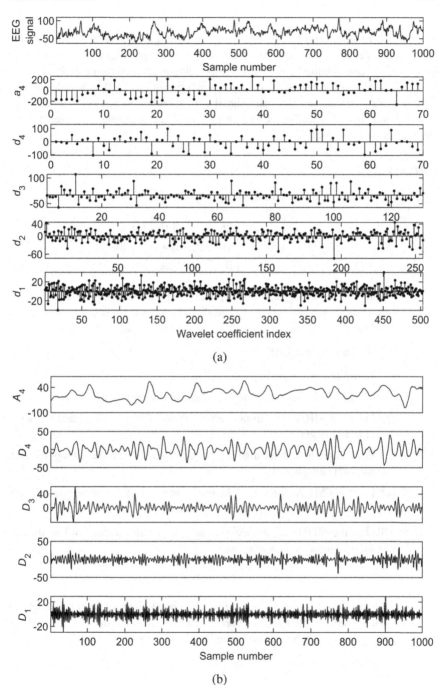

**Figure 5.17** (a) EEG signal and DWT coefficients for 4-level decomposition and (b) 4-level DWT based sub-band signals.

has been sampled at 160 Hz. In Fig. 5.17 (a), $a_l$ and $d_l$ denote the level-$l$ approximation and detail coefficients, respectively. The sub-band signals $A_l$ and $D_l$ at the level-$l$ have been obtained from the corresponding approximation and detail coefficients $a_l$ and $d_l$ at the level-$l$, which are shown in Fig. 5.17 (b). It should be noted that in order to obtain the sub-band signal corresponding to a specific detail or approximation coefficients all other remaining coefficients are replaced by zeros during the reconstruction process.

## PROBLEMS

Q 5.1  Consider a triangular function $x_T(t)$ defined as,

$$
x_T(t) = \begin{cases} 2t, & \text{if } 0 \leq t < 4 \\ 16 - 2t, & \text{if } 4 \leq t \leq 8 \\ 0, & \text{otherwise} \end{cases}
$$

Compute the scalogram of the signal $x_T(t)$ by considering any wavelet function based on your suitability and discuss about the resolution around time instant $t = 4$. Consider the suitable sampling frequency for simulation.

Q 5.2  Consider the signal $x(t) = e^{-5t}u(t)$.

(a) Verify that the signal $x(t) \in L_2(\mathbb{R})$.

(b) For Haar MRA, determine the projection of $x(t)$ on the $V_0$ space.

(c) For Haar MRA, determine the projection of $x(t)$ on the space $V_m$, where $m$ is an integer.

(d) For Haar MRA, determine the projection of $x(t)$ on the orthogonal complement subspace $W_0$.

(e) For Haar MRA, determine the projection of $x(t)$ on the orthogonal complement subspace $W_m$, where $m$ is an integer.

Q 5.3  Sketch the STFT-based spectrogram and wavelet transform-based scalogram of a signal $x(t)$ defined as,

$$
x(t) = \cos(\omega_1 t) + \cos(\omega_2 t)
$$

where, $\omega_1 < \omega_2$. Also comment on the resolutions of both the frequency components in the obtained spectrogram and scalogram. Consider the suitable frequencies $\omega_1$, $\omega_2$ and sampling frequency for simulation.

Q 5.4  Consider the signal $x(t)$ defined as,

$$
x(t) = \frac{1}{\sqrt{\alpha}} \psi \left( \frac{t - n\alpha}{\alpha} \right)
$$

where $\psi(t)$ is Haar wavelet. Sketch the signal $x(t)$ for all possible combinations of scale $\alpha$ ($\alpha = 2^{-2}, 2^{-1}, 1, 2, 4$) and translation parameter $n$ ($n = -2, -1, 0, 1, 2$). Also, comment on the orthogonality of these scaled and translated signals.

Q 5.5 Prove the properties of the CWT mentioned in Table 5.1.

Q 5.6 Consider the signal $x(t)$ defined as,

$$x(t) = \cos(\omega_1 t) + 0.005\delta(t - t_1) + 0.005\delta(t - t_2), \quad t_1 < t_2$$

Study the CWT coefficients at different scales for discontinuity analysis. Also, discuss the energy distributions at different scales. Consider suitable parameters $\omega_1, t_1, t_2$, and sampling frequency for simulation. Also, $\delta(t)$ can be considered as Kronecker delta for simulation.

Q 5.7 Three-level dyadic and irregular tree-based filter banks are shown in Fig. 5.18. Sketch the time-frequency tilling patterns corresponding to all the filter banks. Note that L represents low-frequency band and H represents high-frequency band of the signal.

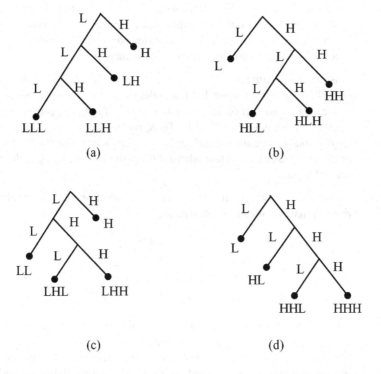

**Figure 5.18** (a)-(d) Three-level dyadic and irregular tree-based filter banks for Q 5.7.

Q 5.8 Show the scalogram of the following signals:

(a) $x_1(t) = \cos(2\pi 100t + 20\pi t^2)$

(b) $x_2(t) = \cos(2\pi 300t + 50\pi t^3)$

(c) $x_3(t) = x_1(t) + x_2(t)$

(d) $x_4(t) = \cos(2\pi 100t + 20\sin(5\pi t))$

Select suitable sampling frequency for simulation.

Q 5.9 Consider the signal $x(t) = \cos[2\pi(140 + 70t)t]$, for simulation choose suitable sampling frequency. Add additive white Gaussian noise with 20 dB signal to noise ratio (SNR) to the sampled signal in order to have noisy signal. Perform the 6-level DWT and determine the energy of the DWT coefficients at each level. Also, plot the distribution of energy of coefficients at different levels, based on this distribution determine the significant coefficients. Then, perform inverse DWT of the significant coefficients to obtain the DWT-based denoised signal. Compute the MSE of the denoised and noisy signals.

Q 5.10 Consider the impulse signal $x(t) = \delta(t)$, compute the DWT of the signal $x(t)$ up to three-level. Now, reconstruct the signal using obtained coefficients at each level and compute its Fourier transform to analyze resolution of wavelet transform at each level. Consider $\delta(t)$ as a Kronecker delta and suitable sampling frequency for simulation purpose.

Q 5.11 Consider a real-time EEG/ECG signal $x(n)$ of length 512 samples. Perform DWT at 5-level using Haar wavelet system to obtain approximation coefficients $a_5$ and set of detail coefficients $d_5$, $d_4$, $d_3$, $d_2$, and $d_1$. First, reconstruct the signal using $a_5$ only. Then, reconstruct using $a_5$ and $d_5$ combinedly and so on until we include all the coefficients. Compute the set of coefficients required to retain atleast 90% of the signal energy in the reconstructed signal.

Q 5.12 Consider a piecewise-constant signal $x(t)$ defined over the interval $[0, 1]$, which is mathematically expressed as,

$$x(t) = \begin{cases} 2, & \text{if } 0 \le t \le \frac{1}{4} \\ 0, & \text{if } \frac{1}{4} \le t \le \frac{1}{2} \\ 4, & \text{if } \frac{1}{2} \le t \le \frac{3}{4} \\ -2, & \text{if } \frac{3}{4} \le t \le 1 \end{cases}$$

Find out the Haar wavelet transform of the signal $x(t)$ using the basis functions $\phi_{0,0}(t)$, $\psi_{0,0}(t)$, $\psi_{-1,0}(t)$, and $\psi_{-1,1}(t)$. Also, find out the reconstructed signal using the inverse Haar wavelet transform. Sketch and compare the original signal $x(t)$ and reconstructed signal. Note $\phi(t)$ and $\psi(t)$ are the scaling function and mother wavelet for Haar wavelet transform, respectively.

# 6 Quadratic Time-Frequency Transforms

*"Tell me and I forget. Teach me and I remember. Involve me and I learn."* –Benjamin Franklin

## 6.1 QUADRATIC TIME-FREQUENCY TRANSFORMS

It should be noted that the STFT and wavelet transform-based TFA depend on the window function or mother wavelet. In the case of STFT, this window function is fixed, whereas in wavelet transform, it is scale-dependent or can be adapted according to the requirement. There are time-frequency transforms which do not need the window function for obtaining TFR [61]. Such transforms are known as quadratic time-frequency transforms, the WVD and ambiguity function are main parts of this transform family. One of the important advantages of the quadratic time-frequency transform is that it does not require the window function for analysis, so selection of the window function is not needed for transformation. Moreover, such transforms do not require design of basis functions for TFR. In the case of STFT and wavelet transform, the spectral representation of the signal depends on the type and size of the window or wavelet function. But, such type of problem is not present in the quadratic TFR. Due to the absence of window function or wavelet in the quadratic time-frequency transforms, these transforms provide better estimation of time-varying frequency and fulfill the marginal conditions. But such type of time-frequency transforms suffer from a major limitation which is the presence of cross-term specially in case of nonlinear chirp signals or multi-component signals which may mislead the interpretation of TFR. These quadratic time-frequency transforms include several TFA methods. The WVD has been discussed in the next section.

### 6.1.1 WIGNER-VILLE DISTRIBUTION

The WVD is the oldest TFR and considered as a classical time-frequency transform method. Theoretically, it has infinite resolution in both time-domain and frequency-domain [62, 63]. This TFR has got many applications in signal analysis and computer vision. One of the major difficulties with this transform is the presence of cross-term in the analysis of nonlinear chirp signals or multi-component non-stationary signals. The cross-term existing between the components are termed as inter cross-term and the cross terms present due to the nonlinear frequency modulation are termed as intra cross-term [64].

DOI: 10.1201/9781003367987-6

The WVD can have two forms known as cross WVD and auto WVD. The cross WVD of the signal is defined as [65, 66, 67],

$$\text{WVD}_{x,y}(t,\omega) = \int_{-\infty}^{\infty} x\left(t+\frac{\tau}{2}\right) y^*\left(t-\frac{\tau}{2}\right) e^{-j\omega\tau} d\tau \tag{6.1}$$

In the above expression, the WVD can be considered as Fourier transform of the time-varying cross correlation function and in this way WVD can be interpreted as a time-varying power spectral density (PSD). The auto WVD for signal $x(t)$ can be expressed as,

$$\text{WVD}_x(t,\omega) = \int_{-\infty}^{\infty} x\left(t+\frac{\tau}{2}\right) x^*\left(t-\frac{\tau}{2}\right) e^{-j\omega\tau} d\tau \tag{6.2}$$

In frequency-domain, the auto WVD for signal $x(t)$ is defined as,

$$\text{WVD}_x(t,\omega) = \frac{1}{2\pi} \int_{-\infty}^{\infty} X\left(\omega-\frac{\zeta}{2}\right) X^*\left(\omega+\frac{\zeta}{2}\right) e^{-jt\zeta} d\zeta$$

It should be noted that WVD is not non-negative or positive for all the signals. In some cases, WVD can take negative values. In spite of it, WVD has very desirable properties and it is useful for designing other TFDs and variants of WVD. Now, we will see some examples for computing WVD.

### 6.1.1.1   WVD of an Impulse Function

The WVD of the impulse function, $x(t) = A\delta(t-T)$ can be computed by Eq. (6.2),

$$\text{WVD}_x(t,\omega) = \int_{-\infty}^{\infty} A\delta\left(t+\frac{\tau}{2}-T\right) A\delta^*\left(t-\frac{\tau}{2}-T\right) e^{-j\omega\tau} d\tau$$

Putting, $t-\tau/2-T = 0$ and $dt = d\tau/2$

We have,

$$\text{WVD}_x(t,\omega) = \int_{-\infty}^{\infty} A^2\delta(\tau)\delta(0)e^{-j\omega(t-T)2}\, 2\, dt$$

$$= 2\,A^2\delta(\tau)e^{-j\omega(t-T)2}$$

$$= A^2\delta(t-T)e^{-j\omega(t-T)2}$$

It should be clear that at point $t = T$ all frequencies are present in the time-frequency domain. Plots of impulse signal $x(t) = \delta(t-10)$ and its WVD are shown in Fig. 6.1. For simulation, Kronecker delta function has been used to represent $\delta(t)$ and suitable sampling frequency has been considered.

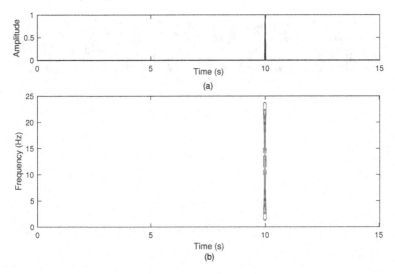

**Figure 6.1** (a) An impulse signal and (b) its WVD-based TFR.

### 6.1.1.2 WVD of a Complex Exponential Signal

The WVD of the complex exponential signal can be computed using Eq. (6.2) as follows:

$$x(t) = Be^{j\omega_k t}, \text{ where } \omega_k = k\omega_0,$$

$$\text{WVD}_x(t, \omega) = \int_{-\infty}^{\infty} Be^{j\omega_k\left(t+\frac{\tau}{2}\right)} Be^{-j\omega_k\left(t-\frac{\tau}{2}\right)} e^{-j\omega\tau} d\tau$$

$$= B^2 \int_{-\infty}^{\infty} e^{-j\tau(\omega-\omega_k)} d\tau$$

$$= B^2 2\pi\delta(\omega - \omega_k) = 2\pi B^2 \delta(\omega - k\omega_0)$$

It shows that WVD of a complex exponential signal of frequency $\omega_k$ is present at all time points in time-frequency domain. The real and imaginary parts of a complex exponential signal $x(t) = e^{j100\pi t}$ and its WVD are shown in Fig. 6.2. For simulation purpose, the signal $x(t)$ is sampled at sampling rate of 500 Hz.

### 6.1.1.3 WVD of a Chirp Signal

The WVD computation of a chirp signal $x(t) = Be^{j(Ct^2+D)}$, can be computed for Eq. (6.2) as follows:

$$\text{WVD}_x(t, \omega) = \int_{-\infty}^{\infty} Be^{j\left[C\left(t+\frac{\tau}{2}\right)^2+D\right]} Be^{-j\left[C\left(t-\frac{\tau}{2}\right)^2+D\right]} e^{-j\omega\tau} d\tau$$

$$= B^2 \int_{-\infty}^{\infty} e^{-j\tau(\omega-2Ct)} d\tau$$

$$= 2\pi B^2 \delta(\omega - 2Ct)$$

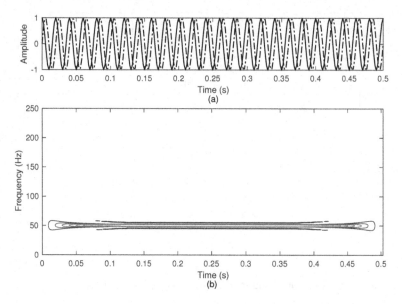

**Figure 6.2** (a) Real and imaginary parts of a complex exponential $x(t) = e^{j100\pi t}$ are shown by solid and dash-dot lines, respectively, and (b) WVD-based TFR.

Here $2Ct$ is IF of the considered chirp signal $x(t)$. The WVD plot provides significant values at IF only. Figure 6.3 shows a chirp signal $x(t) = e^{j2\pi(100t^2+5)}$ and its WVD plot. For simulation, the chirp signal $x(t)$ is sampled at sampling frequency of 500 Hz.

In the examples, complex signals instead of real signals have been considered. The reason for considering complex signal is that any real signal can be written as a sum of complex signals corresponding to positive frequency parts and negative frequency parts and both parts together can produce cross terms. In practice, before computing WVD of real signals, we convert them into analytic signals (or complex signals) in order to avoid cross terms obtained due to the positive and negative frequency components present in the signals.

### 6.1.2  PROPERTIES OF WVD

WVD has interesting properties which are listed in Table 6.1.

### 6.2  CROSS-TERM SUPPRESSION IN WVD

One of the major concerns in WVD analysis is the presence of cross-term in case of multi-component signal. For reducing the effect of cross-term in TFR obtained from WVD, various methods like, kernel-based approaches, signal decomposition-based techniques, etc. have been proposed in the literature.

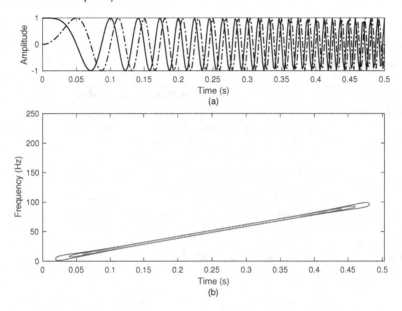

**Figure 6.3** Real and imaginary parts of a chirp signal $x(t) = e^{j2\pi(100t^2+5)}$ shown by solid and dash-dot lines, respectively, in (a) and its corresponding WVD-based TFR in (b).

The Choi-Williams distribution is a kernel-based approach to resolve the problem of cross-term in WVD. In this method, exponential kernel is used. The pseudo-WVD (PWVD) is a window-based approach to resolve the issue of cross-term in WVD. The smoothed PWVD (SPWVD) was proposed to improve the performance of PWVD by using an additional filter to suppress the cross-term in both time and frequency directions.

There are various signal decomposition-based techniques to suppress the cross-term in WVD like, tunable-Q wavelet transform (TQWT) based technique [68], emprical wavelet transform (EWT) based technique [69], empirical mode decomposition (EMD) based technique [70], iterative eigenvalue decomposition of Hankel matrix (IEVDHM) based technique [71], sliding eigenvalue decomposition (SEVD) based technique [72], variational mode decomposition (VMD) based technique [73], FBSE-based technique [74], Gabor expansion-based technique [75], perfect reconstruction filter bank (PRFB) based technique [76, 77], sliding mode singular spectrum analysis based approach [78], etc. In the next subsection, the FBSE-based approach for cross-term reduction has been explained.

### 6.2.1 FBSE-BASED TECHNIQUE TO SUPPRESS CROSS-TERM

In this technique, the idea is to decompose a multi-component signal into monocomponent signals and then applying WVD on analytic representation of each of the

**Table 6.1**
**Properties of WVD**

| Property | Mathematical Expression |
|---|---|
| Real valued | $\text{WVD}_x^*(t,\omega) = \text{WVD}_x(t,\omega)$ |
| Symmetry | For real signal symmetric spectrum, i.e., $X(\omega) = X(-\omega)$: $\text{WVD}_x(t,\omega) = \text{WVD}_x(t,-\omega)$ <br><br> For real spectrum symmetric signal, i.e., $x(t) = x(-t)$: $\text{WVD}_x(t,\omega) = \text{WVD}_x(-t,\omega)$ |
| Marginals and energy | Time marginal: $\int_{-\infty}^{\infty} \text{WVD}_x(t,\omega)d\omega = \|x(t)\|^2$ <br><br> Frequency marginal: $\int_{-\infty}^{\infty} \text{WVD}_x(t,\omega)dt = \|X(\omega)\|^2$ <br><br> Energy$= \int_{-\infty}^{\infty}\int_{-\infty}^{\infty} \text{WVD}_x(t,\omega)dt\,d\omega$ <br> $= \int_{-\infty}^{\infty} \|X(\omega)\|^2 d\omega = \int_{-\infty}^{\infty} \|x(t)\|^2 dt$ |
| Time and frequency shifts | If $y(t) = e^{j\omega_0 t}x(t-t_0)$ then, $\text{WVD}_y(t,\omega) = \text{WVD}_x(t-t_0, \omega-\omega_0)$ |
| Non-zero time and non-zero frequency limited signals | If $x(t) = 0$ for $t < t_1$ and/or $t > t_2$ then, $\text{WVD}_x(t,\omega) = 0$ for $t < t_1$ and/or $t > t_2$ <br><br> Similarly, <br> If $X(\omega) = 0$ for $\omega < \omega_1$ and/or $\omega > \omega_2$ then, $\text{WVD}_x(t,\omega) = 0$ for $\omega < \omega_1$ and/or $\omega > \omega_2$ |
| Mean time, mean frequency, duration, and bandwidth | The computation of these quantities depends on the marginals, which can be obtained correctly for WVD. |
| Non-linearity | If $x(t) = a_1 x_1(t) + a_2 x_2(t)$ then, $\text{WVD}_x(t,\omega) = \underbrace{a_1^2\,\text{WVD}_{x_1}(t,\omega) + a_2^2\,\text{WVD}_{x_2}(t,\omega)}_{\text{Auto-term}}$ $+ \underbrace{2\text{Re}(a_1 a_2^* \text{WVD}_{x_1,x_2}(t,\omega))}_{\text{Cross-term}}$ |
| Dilation | If $x_1(t) = \frac{1}{\sqrt{a}}x\left(\frac{t}{a}\right)$, $a \neq 0$ then, $\text{WVD}_{x_1}(t,\omega) = \text{WVD}_x\left(\frac{t}{a}, a\omega\right)$ |
| Time scaling | If $x_1(t) = x(at)$ then, $\text{WVD}_{x_1}(t,\omega) = \frac{1}{\|a\|}\text{WVD}_x\left(at, \frac{\omega}{a}\right)$ |

**Table 6.1**

**Properties of WVD (Continued)**

| Property | Mathematical Expression |
|---|---|
| Signal reconstruction | Inverse Fourier transform of $\text{WVD}_x(t,\omega)$ with respect to $\omega$ can be given by, $$x\left(t+\tfrac{\tau}{2}\right)x^*\left(t-\tfrac{\tau}{2}\right) = \tfrac{1}{2\pi}\int_{-\infty}^{\infty} \text{WVD}_x(t,\omega)e^{j\omega t}\,d\omega$$ at $t=\tfrac{\tau}{2}$, reconstructed signal can be obtained as, $x^*(0)x(\tau)$. Hence, the signal $x(t)$ can be reconstructed with a multiplication factor $x^*(0)$. |
| Multiplication in time | If $y(t) = x_1(t)x_2(t)$ then, $$\text{WVD}_y(t,\omega) = \tfrac{1}{2\pi}\text{WVD}_{x_1}(t,\omega) * \text{WVD}_{x_2}(t,\omega)$$ $$= \tfrac{1}{2\pi}\int_{-\infty}^{\infty} \text{WVD}_{x_1}(t,\omega_1)\text{WVD}_{x_2}(t,\omega-\omega_1)d\omega_1$$ Here, $*$ denotes convolution operation with respect to frequency. |
| Convolution in time | If $y(t) = x_1(t) * x_2(t)$ then, $$\text{WVD}_y(t,\omega) = \text{WVD}_{x_1}(t,\omega) * \text{WVD}_{x_2}(t,\omega)$$ $$= \int_{-\infty}^{\infty} \text{WVD}_{x_1}(t_1,\omega)\text{WVD}_{x_2}(t-t_1,\omega)dt_1$$ In the above expression, the convolution operation with respect to time is represented by $*$. |

decomposed components [74]. To obtain the final TFR, all the TFRs obtained from each of the components are used. The decomposition of multi-component signal $x(t)$ into mono-component signals using order-zero FBSE is performed by representing $x(t)$ in terms of FBSE coefficients $a_k$ and then, performing inverse FBSE of various groups of the FBSE coefficients. The signal $x(t)$ can be expressed in terms of the separated mono-component signals $x_1(t), x_2(t), \ldots, x_K(t)$. The WVD of $i^{\text{th}}$ separated component $x_i(t)$ using Eq. (6.2) is expressed as,

$$\text{WVD}_{z_{x_i}}(t,\omega) = \int_{-\infty}^{\infty} z_{x_i}\left(t+\frac{\tau}{2}\right)z_{x_i}^*\left(t-\frac{\tau}{2}\right)e^{-j\omega\tau}d\tau, \quad i=1,2,\ldots,K \qquad (6.3)$$

where $z_{x_i}(t)$ is analytic representation of the signal $x_i(t)$ such that, $z_{x_i}(t) = x_i(t) + H\{x_i(t)\}$ as shown in Eq. (3.64).

Then, WVDs of each analytic component are added together to obtain cross-term free WVD of signal $x(t)$ as,

$$\text{WVD}_x(t,\omega) = \sum_{i=1}^{K} \text{WVD}_{z_{x_i}}(t,\omega) \qquad (6.4)$$

For simulation study, a multi-component signal $x(n)$ has been considered which is expressed by Eq. (6.5) and shown in Fig. 6.4 (a). Figures 6.4 (b) and (c) show the

WVD with the presence of cross-term and the WVD with cross-term suppression using FBSE-based technique, respectively.

$$x(n) = \sum_{i=1}^{2} A_i \cos\left[2\pi\left(f_i + \frac{1}{2}\beta_i nT_s\right) nT_s\right] \tag{6.5}$$

where $A_1 = 1.4$, $A_2 = 1$, $f_1 = 40$, $f_2 = 310$, $\beta_1 = 90$, $\beta_2 = -60$, $n = 0, 1, \ldots, 1000$, and $T_s = \frac{1}{F_s} = 1$ ms. Here, $F_s$ represents the sampling rate of the signal.

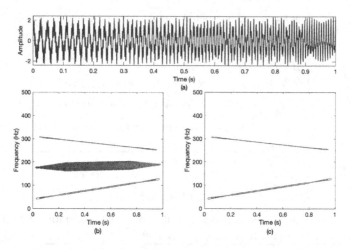

**Figure 6.4**　(a) Multi-component signal $x(n)$ in time-domain, (b) the WVD-based TFR for signal $x(n)$, and (c) cross-term free TFR based on the WVD of signal $x(n)$ obtained using FBSE-based technique.

### 6.2.2　PSEUDO WVD

The PWVD of a signal $x(t)$ is defined as windowed WVD of $x(t)$ [79] which can be mathematically expressed as,

$$\text{PWVD}_x(t,\omega) = \int w(\tau) x\left(t + \frac{\tau}{2}\right) x^*\left(t - \frac{\tau}{2}\right) e^{-j\omega\tau} d\tau \tag{6.6}$$

The PWVD is suitable for online data processing as it processes the data of short duration. It should be noted that the windowing in time-domain provides smoothing in frequency-domain, as a result, the PWVD suppresses the cross-term oscillating in frequency direction only. The PWVD can also be expressed as a convolution of WVD and the window function $w(t)$ in Fourier domain ($W(\omega)$) as,

$$\text{PWVD}_x(t,\omega) = \frac{1}{2\pi} \text{WVD}_x(t,\omega) * W(\omega) \tag{6.7}$$

where $*$ represents the convolution operation.

For an example, a non-stationary signal $x(n)$ with $n = 0, 1, \ldots, 1000$ has been considered. The mathematical expression of $x(n)$ is shown in expression below.

$$x(n) = \sum_{i=1}^{2} h(n)A_i \cos(2\pi f_i n T_s) \tag{6.8}$$

where $A_1 = 0.9$, $A_2 = 0.74$, $f_1 = 150$ Hz, $f_2 = 400$ Hz, $T_s = 0.001$ s, and $h(n)$ is expressed as,

$$h(n) = \begin{cases} 0, & \text{if } 0 \leq n \leq 150 \\ w(n), & \text{if } 151 \leq n \leq 300 \\ 0, & \text{if } 301 \leq n \leq 700 \\ w(n), & \text{if } 701 \leq n \leq 850 \\ 0, & \text{otherwise} \end{cases} \tag{6.9}$$

where $w(n)$ is the Hamming window of length 150. The signal $x(n)$ has been shown in Fig. 6.5 (a). The WVD- and PWVD-based TFRs of the signal $x(n)$ are shown in Figs. 6.5 (b) and (c), respectively.

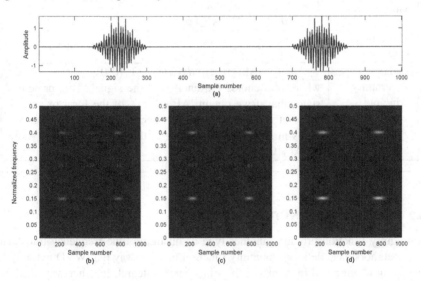

**Figure 6.5** (a) Signal $x(n)$, (b) WVD-based TFR, (c) PWVD-based TFR, and (d) SPWVD-based TFR for signal $x(n)$.

### 6.2.3 SMOOTHED PSEUDO WVD

In order to apply smoothing in time direction, the whole expression of PWVD can be convolved with a low-pass filter $g(u)$ [80] which can be mathematically expressed as,

$$SPWVD_x(t, \omega) = \int g(u - t) \int w(\tau) x \left( u + \frac{\tau}{2} \right) x^* \left( u - \frac{\tau}{2} \right) e^{-j\omega\tau} d\tau du \tag{6.10}$$

Here, the window function $w(t)$ and low-pass filter $g(t)$ are independent of each other, which makes the SPWVD very versatile in cross-term suppression. Also, the independency between $w(t)$ and $g(t)$ helps in selecting both the functions based on the amount of smoothing needed in both the directions, i.e., frequency and time, respectively [79]. Furthermore, SPWVD removes the cross-term in both time and frequency directions which was not the case in PWVD. For the signal $x(n)$ which is represented mathematically by Eq. (6.8) and shown in Fig. 6.5 (a), the SPWVD has been shown in Fig. 6.5 (d).

**Table 6.2**
**Properties of Ambiguity Function**

| Property | Mathematical Expression |
|---|---|
| Time shift | If $x_1(t) = x(t-t_0)$ then,<br>$AF_{x_1}(\vartheta,\tau) = e^{j\vartheta t_0} AF_x(\vartheta,\tau)$ |
| Modulation | If $x_1(t) = e^{j\omega_0 t}x(t)$ then,<br>$AF_{x_1}(\vartheta,\tau) = e^{j\omega_0 \tau} AF_x(\vartheta,\tau)$ |
| Maximum value | $\max\{AF_x(\vartheta,\tau)\} = AF_x(0,0) = $ Energy of the signal |
| Radar uncertainty principle | $\frac{1}{2\pi}\int_{-\infty}^{\infty}\int_{-\infty}^{\infty}|AF_x(\vartheta,\tau)|^2 d\tau d\vartheta = |AF_x(0,0)|^2 = E^2$<br>where, $E$ denotes the energy of the signal. This principle states that if we obtain $|AF_x(\vartheta,\tau)|^2$ of the form of an impulse at the origin then, it is necessary to cover other regions in the $(\tau,\vartheta)$ plane due to the limited maximal value of $|AF_x(0,0)|^2 = E^2$. |

### 6.2.4  AMBIGUITY FUNCTION

Ambiguity function provides another type of quadratic time-frequency transform. Some authors have defined ambiguity function in same way like WVD but only by swapping of time and integral variables in Fourier integral. In other way, the term $e^{-j\vartheta t}$ has been considered as $e^{j\vartheta t}$, where $\vartheta$ is replaced by $-\vartheta$. Here, the second convention is considered for presenting the concepts in following sections. The ambiguity function for signal $x(t)$ is defined as,

$$AF_x(\vartheta,\tau) = \int_{-\infty}^{\infty} x\left(t+\frac{\tau}{2}\right)x^*\left(t-\frac{\tau}{2}\right)e^{j\vartheta t}dt$$

Based on Parseval's relation, we can write this expression in frequency-domain as,

$$AF_x(\vartheta,\tau) = \frac{1}{2\pi}\int_{-\infty}^{\infty} X\left(\omega-\frac{\vartheta}{2}\right)X^*\left(\omega+\frac{\vartheta}{2}\right)e^{j\omega\tau}d\omega$$

The ambiguity function has the properties which are shown in Table 6.2. In the similar way, the cross ambiguity function is defined as follows:

$$AF_{x,y}(\vartheta,\tau) = \int_{-\infty}^{\infty} x\left(t+\frac{\tau}{2}\right) y^*\left(t-\frac{\tau}{2}\right) e^{j\vartheta t} dt$$

In frequency-domain, we have,

$$AF_{x,y}(\vartheta,\tau) = \frac{1}{2\pi} \int_{-\infty}^{\infty} X\left(\omega-\frac{\vartheta}{2}\right) Y^*\left(\omega+\frac{\vartheta}{2}\right) e^{j\omega\tau} d\omega$$

## 6.2.5 RELATIONSHIP BETWEEN AMBIGUITY FUNCTION AND WVD

The $WVD_x(t,\omega)$ can be considered as the 2D Fourier transform of $AF_x(\vartheta,\tau)$ [51], which is mathematically expressed as,

$$WVD_x(t,\omega) = \frac{1}{2\pi} \int_{-\infty}^{\infty} \int_{-\infty}^{\infty} AF_x(\vartheta,\tau) e^{-j\vartheta t} e^{-j\omega\tau} d\vartheta d\tau$$

This can also be viewed as performing two subsequent 1D Fourier transforms. The relation between WVD and ambiguity function is shown in Fig. 6.6.

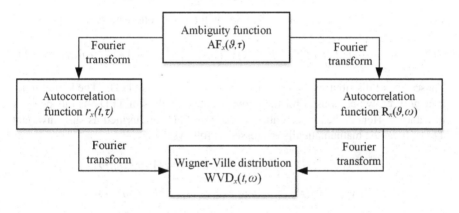

**Figure 6.6**   Relationship between ambiguity function and WVD.

The Fourier transform of ambiguity function with respect to $\vartheta$ can be written as,

$$r_x(t,\tau) = \frac{1}{2\pi} \int_{-\infty}^{\infty} AF_x(\vartheta,\tau) e^{-j\vartheta t} d\vartheta$$
$$= x^*\left(t-\frac{\tau}{2}\right) x\left(t+\frac{\tau}{2}\right)$$

The Fourier transform of $AF_x(\vartheta,\tau)$ with respect to $\tau$ gives,

$$R_x(\vartheta,\omega) = \int_{-\infty}^{\infty} AF_x(\vartheta,\tau) e^{-j\omega\tau} d\tau$$
$$= X\left(\omega-\frac{\vartheta}{2}\right) X^*\left(\omega+\frac{\vartheta}{2}\right)$$

We can have,

$$\mathrm{WVD}_x(t,\omega) = \int_{-\infty}^{\infty} r_x(t,\tau)e^{-j\omega\tau}d\tau = \frac{1}{2\pi}\int_{-\infty}^{\infty} R_x(\vartheta,\omega)e^{-j\vartheta t}d\vartheta$$

It should be noted that WVD can take negative values and cannot be interpreted as a point-wise distribution of energy. Except for this problem, WVD has all the required properties for TFD.

## 6.3   GENERAL TIME-FREQUENCY DISTRIBUTIONS

The WVD works well for linear chirp signals but for general signals, it has negative values and it becomes difficult to consider it as a true density function. The solution for this problem is to use 2D smoothing kernel functions which makes sure that the TFD is positive for all signals. On the other hand, the disadvantage of using smoothing kernels is that it may lose other properties. We will consider shift-invariant and affine TFDs for explanation.

### 6.3.1   SHIFT-INVARIANT TIME-FREQUENCY DISTRIBUTIONS

Cohen proposed a general class of TFDs which has the following representation:

$$\mathrm{TF}_x(t,\omega) = \frac{1}{2\pi}\int\int\int e^{j\vartheta(u-t)}p(\vartheta,\tau)x^*\left(u-\frac{\tau}{2}\right)x\left(u+\frac{\tau}{2}\right)e^{-j\omega\tau}d\vartheta dud\tau \quad (6.11)$$

These type of distributions are also termed as Cohen's class TFDs. The kernel function $p(\vartheta,\tau)$, which is used for the above-mentioned class of TFDs, is independent of $t$ and $\omega$. Hence, the aforementioned class of TFDs are termed as shift-invariant TFDs. It can be mathematically expressed as follows:

$$\text{If } x_1(t) = x(t-t_0), \text{ then, } \mathrm{TF}_{x_1}(t,\omega) = \mathrm{TF}_x(t-t_0,\omega)$$

$$\text{If } x_1(t) = x(t)e^{j\omega_0 t}, \text{ then, } \mathrm{TF}_{x_1}(t,\omega) = \mathrm{TF}_x(t,\omega-\omega_0)$$

Selection of $p(\vartheta,\tau)$ provides various possible shift-invariant TFDs.
The above-mentioned equation of Cohen's class TFD by solving integration with respect to $u$ can be written as,

$$\mathrm{TF}_x(t,\omega) = \frac{1}{2\pi}\int\int p(\vartheta,\tau)\mathrm{AF}_x(\vartheta,\tau)e^{-j\vartheta t}e^{-j\omega\tau}d\vartheta d\tau \quad (6.12)$$

The above expression can be written as convolution of $\mathrm{WVD}_x(t,\omega)$ with $P(t,\omega)$.

$$\mathrm{TF}_x(t,\omega) = \frac{1}{2\pi}\mathrm{WVD}_x(t,\omega)**P(t,\omega) \quad (6.13)$$

where $\mathrm{WVD}_x(t,\omega)$ is WVD of $x(t)$ and $P(t,\omega) = \frac{1}{2\pi}\int\int p(\vartheta,\tau)e^{-j\vartheta t}e^{-j\omega\tau}d\vartheta d\tau$. The product $p(\vartheta,\tau)\,\mathrm{AF}_x(\vartheta,\tau)$ is known as generalized ambiguity function. This expression indicates that the Cohen's class TFDs can be computed as 2D Fourier transform of 2D generalized ambiguity function. It should be noted that, in some cases, the factor $\frac{1}{4\pi^2}$ is used instead of $\frac{1}{2\pi}$ in the definition of Cohen's class TFDs. Based on the selection of different $p(\vartheta,\tau)$, many TFDs have been proposed as a part of this class of TFDs. Some major TFDs of this class include WVD, Margenau-Hill distribution, Kirkwood & Rihaczek distribution, Born-Jordan distribution, Page distribution, Choi-Williams distribution, spectrogram, Zhao-Atlas-Marks distribution, etc. Three examples of kernel function and corresponding TFDs for Cohen's class are given as follows:

### 6.3.1.1 Wigner-Ville Distribution

The kernel, $p(\vartheta,\tau) = 1$ provides the WVD-based TFD. By putting $p(\vartheta,\tau) = 1$ in Eq. (6.11) and rearranging, we get,

$$\text{TF}_x(t,\omega) = \frac{1}{2\pi} \int \int \left[ \int e^{j\vartheta(u-t)} d\vartheta \right] x^* \left(u - \frac{\tau}{2}\right) x \left(u + \frac{\tau}{2}\right) e^{-j\omega\tau} du d\tau$$

$$= \int \int \delta(u-t) x^* \left(u - \frac{\tau}{2}\right) x \left(u + \frac{\tau}{2}\right) e^{-j\omega\tau} du d\tau$$

$$= \int x^* \left(t - \frac{\tau}{2}\right) x \left(t + \frac{\tau}{2}\right) e^{-j\omega\tau} d\tau$$

### 6.3.1.2 Choi-Williams Distribution

In Eq. (6.11), the kernel function $p(\vartheta,\tau) = e^{\frac{-\vartheta^2\tau^2}{4\pi^2\sigma}}$, $\sigma > 0$ is used in order to obtain Choi-Williams distribution [81] which can be expressed as follows:

$$\text{TF}_x(t,\omega) = \frac{1}{2\pi} \int \int \int e^{j\vartheta(u-t)} e^{\frac{-\vartheta^2\tau^2}{4\pi^2\sigma}} x^* \left(u - \frac{\tau}{2}\right) x \left(u + \frac{\tau}{2}\right) e^{-j\omega\tau} d\vartheta du d\tau$$

$$= \frac{1}{2\pi} \int \int \left[ \int e^{j\vartheta(u-t)} e^{\frac{-\vartheta^2\tau^2}{4\pi^2\sigma}} d\vartheta \right] x^* \left(u - \frac{\tau}{2}\right) x \left(u + \frac{\tau}{2}\right) e^{-j\omega\tau} du d\tau$$

$$= \frac{1}{2\pi} \int \int \left[ 2\pi \sqrt{\frac{\pi\sigma}{\tau^2}} e^{\frac{-\pi^2\sigma(u-t)^2}{\tau^2}} \right] x^* \left(u - \frac{\tau}{2}\right) x \left(u + \frac{\tau}{2}\right) e^{-j\omega\tau} du d\tau$$

$$= \int \int \sqrt{\frac{\pi\sigma}{\tau^2}} e^{\frac{-\pi^2\sigma(u-t)^2}{\tau^2}} x^* \left(u - \frac{\tau}{2}\right) x \left(u + \frac{\tau}{2}\right) e^{-j\omega\tau} du d\tau$$

By rearranging the above equation, we have,

$$= \int \int \sqrt{\frac{\pi\sigma}{\tau^2}} e^{-\pi^2\sigma \frac{(u-t)^2}{\tau^2} - j\omega\tau} x^* \left(u - \frac{\tau}{2}\right) x \left(u + \frac{\tau}{2}\right) du d\tau$$

### 6.3.1.3 Spectrogram

In order to obtain spectrogram using Cohen's class TFD, the kernel function $p(\vartheta,\tau) = \int p_1^* \left(t - \frac{\tau}{2}\right) p_1 \left(t + \frac{\tau}{2}\right) e^{j\vartheta t} dt$ is needed. Where $p_1(t)$ is a symmetric window.

The spectrogram of the signal $x(t)$ using window $p_1(t)$ can be computed with the help of Eq. (4.1), as follows:

$$|X(t,\omega)|^2 = \left| \int x(t') p_1(t-t') e^{-j\omega t'} dt' \right|^2$$

If we consider, $x_t(t')$ as $x(t') p_1(t-t')$ and $X_t(\omega)$ is its Fourier transform, then the WVD of $x_t(t')$ and $|X_t(\omega)|^2$ can be related with the help of frequency marginal property as follows:

$$|X_t(\omega)|^2 = \int \text{WVD}_{x_t}(t',\omega) dt'$$

By applying property stated for multiplication of two signals in time-domain equivalent to convolution of their WVDs in frequency-domain and time shifting property of WVD, we have,

$$|X(t,\omega)|^2 = \frac{1}{2\pi} \int \int \text{WVD}_x(t',\omega') \text{WVD}_{p_1}(t-t',\omega-\omega') dt' d\omega'$$

$$= \frac{1}{2\pi} \text{WVD}_x(t,\omega) * * \text{WVD}_{p_1}(t,\omega)$$

Now from Eq. (6.13), we can write,

$$\text{WVD}_{p_1}(t,\omega) = \frac{1}{2\pi} \int \int p(\vartheta,\tau) e^{-j\vartheta t} e^{-j\omega\tau} d\vartheta d\tau$$

From the above equation, we can write,

$$p(\vartheta,\tau) = \int p_1^* \left( t - \frac{\tau}{2} \right) p_1 \left( t + \frac{\tau}{2} \right) e^{j\vartheta t} dt$$

Hence, we can state that the kernel function $p(\vartheta,\tau) = \int p_1^* \left( t - \frac{\tau}{2} \right) p_1 \left( t + \frac{\tau}{2} \right) e^{j\vartheta t} dt$ can be used in Cohen's class TFD to obtain spectrogram.

### 6.3.2 AFFINE-INVARIANT TFDS

This is an alternative approach to perform smoothing of Cohen's class TFDs [82]. The TFD which belongs to the affine class is invariant with respect to time shift and scaling, i.e.,
If $x_1(t) = \sqrt{|a|} x(a(t-t_0))$, then, $\text{TF}_{x_1}(t,\omega) = \text{TF}_x(a(t-t_0),\omega/a)$.
The TFD which satisfies the above-mentioned condition can be computed from WVD with the affine transform as,

$$\text{TF}_x(t,\omega) = \frac{1}{2\pi} \int \int C \left[ \omega(t_1 - t), \frac{\omega_1}{\omega} \right] \text{WVD}_x(t_1,\omega_1) dt_1 d\omega_1$$

This can be interpreted as the correlation process between WVD and kernel $C$ along time axis. The kernel is scaled by varying $\omega$. The conditions for shift-invariant TFD do not exclude the conditions for affine TFD. Due to this reason, there exist

other TFDs besides the WVD which belong to the shift-invariant Cohen's class as
well as to the affine class, for example, Choi-Williams distribution.

For example, the scalogram which is defined as the squared magnitude of wavelet
transform has been considered. For signal $x(t)$ using $\psi_{a,b}(t)$ daughter wavelets with
$\psi(t)$ as mother wavelet, the expression for scalogram ($|\text{WT}(a,b)|^2$) is expressed in
Eq. (5.2). Moyal's formula provides the relationship between the inner product of
two signals and inner product of their corresponding WVDs. For two signals $x(t)$
and $y(t)$ and their WVDs $\text{WVD}_x(t,\omega)$ and $\text{WVD}_y(t,\omega)$, Moyal's formula can be
mathematically expressed as,

$$\left|\int_{-\infty}^{\infty} x(t)y^*(t)dt\right|^2 = \frac{1}{2\pi}\int\int \text{WVD}_x(t,\omega)\text{WVD}_y(t,\omega)\,dt\,d\omega$$

For scalogram, the Moyal's formula can be applied as [83],

$$|\text{WT}(a,b)|^2 = \frac{1}{2\pi}\int\int \text{WVD}_{\psi_{a,b}}(t_1,\omega_1)\text{WVD}_x(t_1,\omega_1)\,dt_1\,d\omega_1$$

where,

$$\text{WVD}_{\psi_{a,b}}(t_1,\omega_1) = \text{WVD}_\psi\left(\frac{t_1-b}{a},a\omega_1\right)$$

Hence,

$$|\text{WT}(a,b)|^2 = \frac{1}{2\pi}\int\int \text{WVD}_\psi\left(\frac{t_1-b}{a},a\omega_1\right)\text{WVD}_x(t_1,\omega_1)\,dt_1\,d\omega_1$$

Putting $b=t$ and $a=\frac{\omega_c}{\omega}$, we have,

$$\text{TF}_x(t,\omega) = \left|\text{WT}\left(t,\frac{\omega_c}{\omega}\right)\right|^2 \tag{6.14}$$

$$= \frac{1}{2\pi}\int\int \text{WVD}_\psi\left(\frac{\omega}{\omega_c}(t_1-t),\frac{\omega_c}{\omega}\omega_1\right)\text{WVD}_x(t_1,\omega_1)\,dt_1\,d\omega_1 \tag{6.15}$$

Hence, scalogram belongs to affine-invariant TFDs.

## 6.4  IMPLEMENTATION OF COHEN'S CLASS TFDS

Eq. (6.13) can be written as,

$$\text{TF}_x(t,\omega) = \frac{1}{2\pi}\int\int \text{WVD}_x(t_1,\omega_1)P(t-t_1,\omega-\omega_1)dt_1 d\omega_1 \tag{6.16}$$

This means all TFDs of Cohen's class can be computed from the convolution of
the WVD with a 2D impulse response $P(t,\omega)$.

Mainly, the kernel $p(\vartheta,\tau)$ suppresses the interference terms of the ambiguity function in TFD $\text{TF}_x(t,\omega)$. From Eq. (6.16), it is clear that reduction of the interference
terms requires smoothing which leads to reduction of time-frequency resolution. Depending on the type of kernel, some of the desired properties of TFDs are preserved.

The interpretation of Cohen's class from Eqs. (6.13) and (6.16) is a straight forward process. First we integrate over $\vartheta$ as,

$$q(u, \tau) = \frac{1}{2\pi} \int_{-\infty}^{\infty} p(\vartheta, \tau) e^{j\vartheta u} d\vartheta$$

Then, we obtain,

$$TF_x(t, \omega) = \int \int q(u - t, \tau) x^* \left( u - \frac{\tau}{2} \right) x \left( u + \frac{\tau}{2} \right) e^{-j\tau\omega} du d\tau$$

The implementation is depicted in Fig. 6.7.

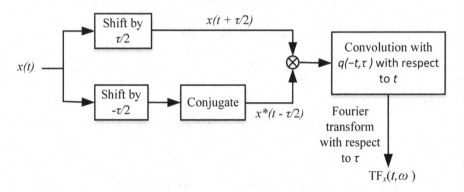

**Figure 6.7**   Implementation of Cohen's class TFDs.

## PROBLEMS

Q 6.1 Compute and sketch the WVD of the following mono-component signals:

    (a) $x(t) = \delta(t - 5)$

    (b) $x(t) = e^{j100\pi t}$

    (c) $x(t) = e^{j2\pi(40+72t)t}$

    (d) $x(t) = \cos(\omega_0 t)$

Q 6.2 Consider a multi-component signal,

$$x(t) = x_0(t - 5)e^{j10\pi t} + x_0(t - 25)e^{j80\pi t}$$

where $x_0(t) = e^{j(100\pi t + 30\pi t^2)}$. Compute the auto-term and cross-term of WVD of the signal $x(t)$. Also discuss the nature of cross-term.

Q 6.3 Compute the PWVD of an complex exponential signal $x(t) = e^{j\omega_0 t}$ with Gaussian window $w(t) = e^{-\frac{5t^2}{2}}$.

Q 6.4 Compute the WVD of the signal,

$$x(t) = e^{j50\pi t} + e^{j300\pi t}$$

Also, compare the resultant WVD with WVD of $e^{j50\pi t}$ and $e^{j300\pi t}$, individually.

Q 6.5 Compute the cross-term in WVD of the following multi-component signals:

(a) $x(t) = \delta(t-5) + \delta(t-15)$

(b) $x(t) = e^{j100\pi t} + e^{j250\pi t}$

Also, compare the obtained cross-term with the simulation results. For simulation, choose suitable sampling frequency and Kronecker delta function to represent $\delta(t)$.

Q 6.6 Prove the properties of WVD mentioned in Table 6.1.

Q 6.7 For a signal,

$$x(t) = e^{j2\pi(30+15t)t} + e^{j2\pi(100+25t)t} + e^{j2\pi(300+40t)t}, \forall t \in [0,1]s.$$

Plot the magnitude spectrum using Fourier transform of the signal. Then, design the filters and separate different components present in the signal. At the end, add the WVD of analytic representation of each of the separated components. Compare the obtained WVD of each separated components and WVD obtained from the signal $x(t)$ directly. Choose appropriate sampling rate for simulation purpose.

Q 6.8 Consider the following four signals:

(a) $x(t) = \cos(100\pi t)$

(b) $x(t) = \sin[2\pi(100+15t)t]$

(c) $x(t) = \cos[2\pi(50+20t^2)t]$

(d) $x(t) = \cos[2\pi(120+15\sin(6\pi t))]$

Compute the WVD of the signals defined over interval 0 to 1 s, directly and by converting into their analytic representation. Discuss the nature of distribution of WVD in both cases for all four signals. Use proper sampling rate for simulation.

Q 6.9 Compute the ambiguity function of the following signal models:

(a) $x(t) = \delta(t-5)$

(b) $x(t) = e^{j100\pi t}$

(c) $x(t) = e^{j2\pi(40+72t)t}$

(d) $x(t) = \cos(\omega_0 t)$

Q 6.10 Prove the properties of ambiguity function mentioned in Table 6.2.

Q 6.11 Consider a LTI system having impulse response $h(t) = \delta(t)$. Find out the WVD of the output signal to this system for input signal $x(t) = e^{j75\pi t}$ using the multiplication property of WVD.

Q 6.12 Find out the WVD of the Gaussian function $x(t) = e^{-\alpha t^2} e^{j\omega_0 t}$.

# 7 Advanced Wavelet Transforms

*"Imagination is more important than knowledge. Knowledge is limited. Imagination encircles the world."* –Albert Einstein

## 7.1 WAVELET PACKET TRANSFORM

The DWT decomposes the signal into a set of different frequency components like Fourier transform but having non-uniform bandwidths for the decomposed components. The DWT can be implemented using filter bank with the help of low-pass and high-pass filters. The output of the low-pass filter provides an approximation component. On the other hand, high-pass filter provides detail component as the output.

Figure 7.1 (a) shows multistage filter bank structure for three-level DWT. The filter bank corresponding to approximation and detail coefficients are shown in Fig. 7.1 (b) where sampling rate is $f_s$ Hz. $d_1(n)$, $d_2(n)$, and $d_3(n)$ are detail coefficients obtained at first, second, and third levels, respectively. Similarly, $c_1(n)$, $c_2(n)$, and $c_3(n)$ are approximation coefficients, obtained at first, second, and third levels, respectively. In Fig. 7.1 (a), LPF and HPF represent low-pass and high-pass filters, respectively.

The detail and approximation coefficients of DWT can be useful for determining various measures for particular band which falls in any of the obtained coefficients from DWT. Sometimes, there is a need of obtaining components corresponding to particular frequency bands which gives the motivations for wavelet packet transform (WPT). Wavelet packets are generalization of wavelet bases, formed by taking linear combinations of usual wavelet functions.

DWT decomposes the approximation spaces $V_m$ only to construct the orthogonal subspaces. On the other hand, WPT decomposes both the approximation space $V_m$ and detail space $W_m$ to construct new bases [48]. Figure 7.2 (a) shows the subspaces associated with WPT. Any node in the tree is labelled by $W_m^p$ represents the subspace for $m^{\text{th}}$ level of decomposition and $p$ $(0 \le p < 2^m)$ is the index of the node at the level $m$. Each node $W_m^p$ has an orthogonal basis $\psi_m^p(t - 2^m n)$, $n \in \mathbb{Z}$. At the root node, $W_0^0 = V_0$, and $\psi_0^0(t) = \phi_0(t)$, where $\phi_0(t)$ is the scaling function for space $V_0$. The scaling function for $V_m$ subspace is defined by $\phi_m(t - 2^m n)$, where $\phi_m(t) = 2^{-m/2}\phi(2^{-m}t)$. By dividing the basis of level $m$, we can construct two orthogonal bases for level $m+1$ with the help of a pair of conjugate mirror filters $h(n)$ and $g(n)$. The relation between the bases of $m^{\text{th}}$ level and $(m+1)^{\text{th}}$ level are given as follows:

$$\psi_{m+1}^{2p}(t) = \sum_{n=-\infty}^{\infty} h(n)\psi_m^p(t - 2^m n) \qquad (7.1)$$

DOI: 10.1201/9781003367987-7

**137**

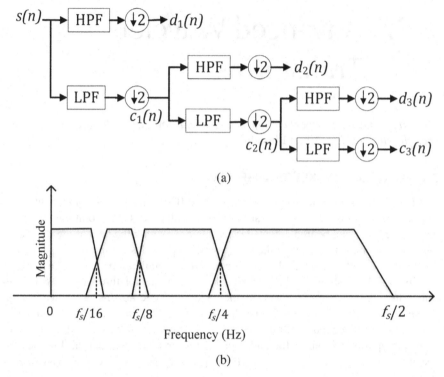

(a)

(b)

**Figure 7.1**  (a) Three-level DWT decomposition tree structure using multistage filter bank and (b) output frequency bands of the three-level DWT.

$$\psi_{m+1}^{2p+1}(t) = \sum_{n=-\infty}^{\infty} g(n)\psi_m^p(t - 2^m n) \qquad (7.2)$$

In DWT, the higher frequency components have poor frequency resolution whereas low-frequency components have good frequency resolution. On the other hand, the WPT can be used to obtain an uniform frequency resolution similar to Fourier decomposition. In WPT, both the detail and approximation coefficients are decomposed to obtain higher level coefficients and this helps to obtain an uniform frequency decomposition [84, 85]. The advantage of WPT is that by proper selection of sampling rate, the frequency bands can be selected to obtain the specific frequency components of the signal from wavelet decomposition trees. The WPT tree for three-level decomposition is shown in Fig. 7.2 (b) for a sampling rate of $f_s$ Hz. The obtained output bands are uniform in nature and shown in Fig. 7.2 (c).

The 'sym4' wavelet has been used for both DWT- and WPT-based TFRs as shown in Fig. 7.3. The WPT is superior in TFA, because WPT divides the frequency axis into finer resolutions than the DWT. To show the superiority of WPT over DWT,

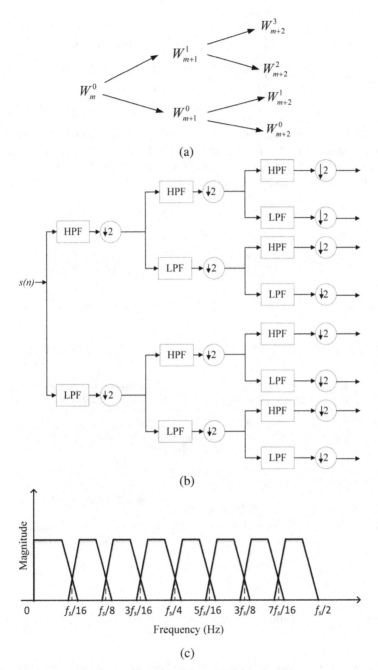

**Figure 7.2** (a) Tree structure for subspaces of three-level WPT, (b) tree structure for three-level WPT, and (c) the output frequency bands of three-level WPT.

the synthetic signal $x(n)$ is considered as,

$$x(n) = \begin{cases} \cos\left(350\pi \frac{n}{f_s}\right), & \text{for } 200 \leq n \leq 400 \\ \cos\left(450\pi \frac{n}{f_s}\right), & \text{for } 600 \leq n \leq 900 \\ 0, & \text{otherwise} \end{cases}$$

where $f_s$ is 1000 Hz.

It can be seen from the Fig. 7.3 that the WPT is able to separate the 175 Hz and 225 Hz components. This is not true for DWT, because 175 Hz and 225 Hz fall in the same band. The bands for a four-level DWT are (in Hz): [0, 31.25), [31.25, 62.5), [62.5, 125), [125, 250), [250, 500).

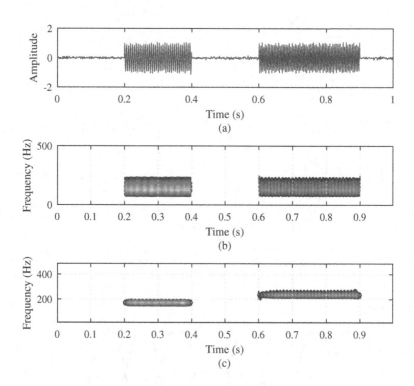

**Figure 7.3** (a) Time-domain signal, (b) TFR based on DWT, and (c) WPT-based TFR.

Figure 7.4 shows two-level decomposed components of EEG signal obtained using WPT where sym3 wavelet function is used. The EEG signal (Espiga3) is obtained from MATLAB. The sampling frequency of the signal is 200 Hz. It should be noted that frequency responses of the output bands depend on the selected wavelet function, chosen decomposition tree, and types of LPF and HPF and the sequence of the filters.

The WPT is also implemented in FBSE domain which has been named as FBSE-based decomposition (FBSED) method [86]. It was used in the literature for automatic discrimination of COVID-19 pneumonia from other viral pneumonia using X-ray and computer tomography (CT) images.

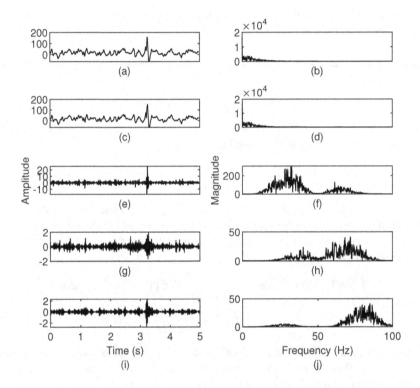

**Figure 7.4**   Two-level decomposition of EEG signal using WPT: (a) EEG signal, (b) spectrum of EEG signal, (c) $d_1(n)$ component, (d) spectrum of $d_1(n)$ component, (e) $d_2(n)$ component, (f) spectrum of $d_2(n)$ component, (g) $d_3(n)$ component, (h) spectrum of $d_3(n)$ component, (i) $d_4(n)$ component, and (j) spectrum of $d_4(n)$ component.

## 7.2   SYNCHROSQUEEZED WAVELET TRANSFORM

The wavelet transform decomposes signal into components by correlating signal with a dictionary of wavelet atoms. The wavelet atoms are translated and scaled version of a mother wavelet. Due to these wavelet atoms, wavelet transform suffers from low time-frequency resolution, i.e., more spread in time-frequency domain.

Synchrosqueezed wavelet transform (SWT) reassigns the signal energy in frequency direction and preserves the time resolution [87]. This reassignment of energy compensates the spreading effect caused by the mother wavelet and since SWT preserves time resolution so inverse SWT can be applied for the reconstruction of the signal. The SWT can characterize the signals having superimposed monocomponents (AFM signal) which are well-separated in the time-frequency domain.

The implementation of SWT can be summarized in three steps as follows:

Step 1: Computation of the wavelet transform $WT(a,b)$ of signal $x(t)$ is carried out using mother wavelet $\psi_{a,b}(t)$ based on Eq. (5.1).

Step 2: Computation of the IF of the signal is performed using the following expression:

$$w_x(a,b) = \frac{\partial}{\partial b}WT(a,b)/(iWT(a,b)) \qquad (7.3)$$

Step 3: Transformation of information from $(a,b)$ plane to the $(WT(a,b),b)$ plane or reallocation of energy is performed as,

$$T_x(b,w_l) = (\Delta w)^{-1} \sum_{a_k:|w_x(a_k,b)-w_l|\leq\Delta w/2} WT(a_k,b)a_k^{-3/2}(\Delta a)_k \qquad (7.4)$$

where $a_k$ is discrete value of the scale $a$ and $\Delta a = a_k - a_{k-1}$. The SWT is only defined at central frequency $w_l$ of the interval $[w_l - (1/2)\Delta w, w_l + (1/2)\Delta w]$ with $\Delta w = w_l - w_{l-1}$. The signal can be reconstructed from the SWT coefficients, $T_x(b, w_l)$ which can be shown as follows:

$$\int_0^\infty WT(a,b)a^{-3/2}da = \frac{1}{2\pi}\int_0^\infty\int_0^\infty X(\xi)\Psi^*(a\xi)e^{ib\xi}a^{-1}dad\xi$$

$$= \int_0^\infty \Psi^*(\xi)\frac{d\xi}{\xi}\frac{1}{2\pi}\int_0^\infty X(\zeta)e^{ib\zeta}d\zeta$$

In the above expression, wavelet coefficients, $WT(a,b)$ are derived by frequency-domain equivalent of Eq. (5.1) obtained using the Parseval's theorem. The signal reconstruction from SWT coefficients can be performed as,

$$x(b) \approx \mathbb{R}\{C_\psi^{-1}\sum_l T_x(b,w_l)(\Delta w)\} \qquad (7.5)$$

where $C_\psi = \frac{1}{2}\int_0^\infty \Psi^*(\xi)\xi^{-1}d\xi$ and $\mathbb{R}\{\cdot\}$ is a real part of complex number.

In order to show the TFRs based on wavelet transform and SWT, the signal $x(n)$ has been used, which can be mathematically expressed as,

$$x(n) = 1.2\cos\left[2\pi\left(250\frac{n}{f_s} + 10\cos\left(4\pi\frac{n}{f_s}\right)\right)\right] \qquad (7.6)$$

where $n = 0,1,...,1000$ and $f_s = 1000$ Hz. The TFRs based on wavelet transform and SWT have been obtained using 'bump' wavelet, which are shown in Fig 7.5.

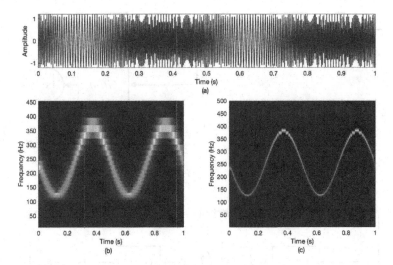

**Figure 7.5** (a) Signal $x(n)$, (b) TFR based on wavelet transform, and (c) SWT-based TFR.

## 7.3 RATIONAL-DILATION WAVELET TRANSFORMS

DWT and WPT provide poor frequency resolution, i.e., low Q-factor (ratio of central frequency to bandwidth) and these transforms do not provide control over selection of Q-factor as they are dyadic type of decomposition [88]. Rational-dilation wavelet transforms (RDWT) can create family of wavelet transform, where Q-factor can be chosen as per requirement demanded by the signal under analysis. It should be noted that the signals with higher oscillatory components require higher value of Q.

The RDWT can be realized with the help of iterated filter bank which is shown in Fig. 7.6 (a) for two-level. Figure 7.6 (b) shows analysis and synthesis filter banks for implementation of RDWT for one-level. The $H(\omega)$ and $G(\omega)$ represent low-pass and high-pass filters in frequency-domain, respectively. Mathematical expression of $H(\omega)$ and $G(\omega)$ can be given as follows:

$$
H(\omega) = \begin{cases} (pq)^{1/2}, & \text{if } \omega \in \left[0, (1-\frac{1}{s})\frac{\pi}{p}\right) \\ (pq)^{1/2}\phi(\frac{\omega-A}{B}), & \text{if } \omega \in \left[(1-\frac{1}{s})\frac{\pi}{p}, \frac{\pi}{q}\right) \\ 0, & \text{if } \omega \in \left[\frac{\pi}{q}, \pi\right] \end{cases} \tag{7.7}
$$

$$
G(\omega) = \begin{cases} 0, & \text{if } \omega \in \left[0, (1-\frac{1}{s})\pi\right) \\ s^{1/2}\phi_c(\frac{\omega-pA}{pB}), & \text{if } \omega \in \left[(1-\frac{1}{s})\pi, \frac{p\pi}{q}\right) \\ s^{1/2}, & \text{if } \omega \in \left[\frac{p\pi}{q}, \pi\right] \end{cases} \tag{7.8}
$$

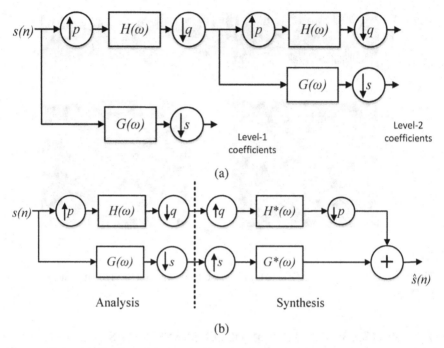

(a)

Analysis                                    Synthesis

(b)

**Figure 7.6** (a) Designed iterated filter bank for implementation of two-level RDWT and (b) analysis and synthesis filter banks for implementation of one-level RDWT.

Condition for perfect reconstruction is given as,

$$\frac{1}{pq}\left|H\left(\frac{\omega}{p}\right)\right|^2 + \frac{1}{s}\left|G(\omega)\right| = 1 \text{ for } \omega \in [0,\pi] \qquad (7.9)$$

Where $A = \left(1 - \frac{1}{s}\right)\frac{\pi}{p}$ and $B = \frac{1}{q} - \left(1 - \frac{1}{s}\right)\frac{1}{p}$, and the values of $A$ and $B$ should be such that $\frac{\omega - A}{B}$ maps $[0,\pi]$ to the transition band of $H(\omega)$. The transition function $\phi(\cdot)$ and complementary transition function $\phi_c(\cdot)$ are defined as follows:

$$\phi(\omega) = 0.5\left[1 + \cos(\omega)\right]\sqrt{2 - \cos(\omega)} \text{ for } \omega \in [0,\pi] \qquad (7.10)$$

and

$$\phi_c(\omega) = \sqrt{1 - \phi^2(\omega)} \qquad (7.11)$$

Parameters $p$, $q$, and $s$ are used for achieving desirable constant Q-factor ($Q$), dilation factor ($d$), and redundancy factor ($r$). Parameters should follow constraints such as, $1 \le p < q$ and $p/q + 1/s \ge 1$. Desired $Q$, $d$, and $r$ of wavelet in terms of $p$, $q$, and $s$ can be obtained using following relations:

$$Q = \frac{\sqrt{pq}}{q - p}, \qquad (7.12)$$

$$d = \frac{q}{p},\qquad(7.13)$$

$$r = \frac{q}{s(q-p)}\qquad(7.14)$$

Figure 7.7 shows an example of RDWT decomposition of ECG signal. ECG signal is taken from MIT-BIH arrhythmia database with sampling rate of 360 Hz [3, 4]. The parameters used for designing the RDWT filter bank are $p = 2$, $q = 3$, $s = 1$, and level of decomposition $= 2$. The TFR obtained for the same ECG signal using HSA on the decomposed components obtained from RDWT is represented in Fig. 7.7.

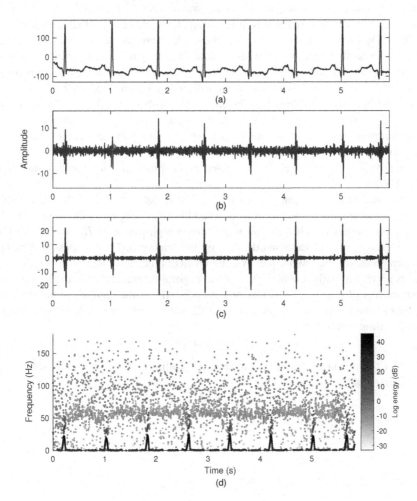

**Figure 7.7**   (a) ECG signal, (b)-(c) decomposed components obtained using RDWT technique, and (d) TFR of the ECG signal.

## 7.4   TUNABLE-Q WAVELET TRANSFORM

The selection of Q-factor for wavelet is required for proper analysis of non-stationary signals. For example, the signals which are oscillatory in nature require high Q-factor for analysis. On the other hand, the signals with low oscillatory components need selection of low Q-factor. Tunable-Q wavelet transform (TQWT) provides flexibility in the of selection of Q-factor [89]. The TQWT is parameterized by Q-factor and oversampling rate (redundancy). The TQWT has been proposed based on the perfect reconstruction oversampled filter bank with real-valued scaling factors. The Q-factor, redundancy factor, and level of decomposition are denoted by $Q$, $r$, and $J$, respectively. The parameter $Q$ adjusts the oscillatory behavior of the wavelet. The parameter $r$ controls excessive ringing so that wavelet is well-localized in time-domain without affecting shape of the wavelet function. It should be noted that the filters used in the TQWT have rational transfer function, which are computationally robust, and can be represented directly in frequency-domain in easy manner. Together with these properties, TQWT follows the perfect reconstruction property of wavelet transform.

The $J^{\text{th}}$ level TQWT-based decomposition can be obtained by applying two-channel filter bank iteratively to the low-pass sub-band coefficients [90]. At every level of decomposition in TQWT, the input signal $s(n)$ of sampling rate $f_s$ is decomposed into two components namely low-pass sub-band coefficients $c_0(n)$ and high-pass sub-band coefficients $d_0(n)$ with sampling rate of $\alpha f_s$ and $\beta f_s$, respectively. The single-level TQWT-based decomposition filter bank is shown in Fig. 7.8 (a).

The output $c_0(n)$ requires a low-pass filter $H_0(\omega)$ which is followed by low-pass scaling operation and has been denoted by LP scaling $\alpha$. In the same way, the high-pass sub-band coefficients $d_0(n)$ requires a high-pass filter $H_1(\omega)$ followed by a high-pass scaling process and denoted by HP scaling $\beta$. The low-pass scaling operation preserves the low-frequency components of the signal and this operation depends on $\alpha$. Similarly, the high-pass scaling operation preserves the high-frequency components of the signal and this process depends on the scaling parameter $\beta$. In order to have perfect reconstruction, the following condition should be satisfied by these scaling parameters:

$$\alpha + \beta > 1 \qquad (7.15)$$

In order to prevent excessive redundancy the scaling parameters should satisfy the following conditions:

$$0 < \alpha < 1 \text{ and } 0 < \beta \leq 1$$

The low-pass scaling is defined as,

$$Y(\omega) = S(\alpha\omega), \ |\omega| \leq \pi$$

and for $0 < \beta \leq 1$, the high-pass scaling is given by,

$$Y(\omega) = \begin{cases} S(\beta\omega + (1-\beta)\pi), & 0 < \omega < \pi \\ S(\beta\omega - (1-\beta)\pi), & -\pi < \omega < 0 \end{cases} \qquad (7.16)$$

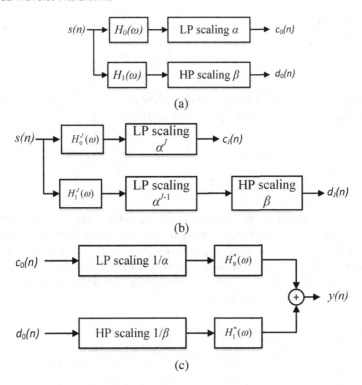

**Figure 7.8** (a) The TQWT-based decomposition based on first-level filter bank, (b) the equivalent system for $J^{\text{th}}$ level TQWT to generate $c_J(n)$ and $d_J(n)$, and (c) the single-level–based reconstruction filter bank for TQWT.

In the above-mentioned expressions, the $S(\omega)$ and $Y(\omega)$ are DTFT of input signal $s(n)$ and output signal $y(n)$ from scaling block, respectively.

The equivalent system corresponding to the $J^{\text{th}}$ level TQWT-based decomposition for input signal $s(n)$ to obtain low-pass sub-band coefficients $c_J(n)$ and high-pass sub-band coefficients $d_J(n)$ have been shown in Fig. 7.8 (b). The equivalent frequency response corresponding to low-pass and high-pass sub-band coefficients obtained after $J^{\text{th}}$ level is given by $H_0^J(\omega)$ and $H_1^J(\omega)$, respectively and which are defined for the case $\alpha, \beta \leq 1$, as follows:

$$H_0^J(\omega) = \begin{cases} \prod_{m=0}^{J-1} H_0\left(\frac{\omega}{\alpha^m}\right), & |\omega| \leq \alpha^J \pi \\ 0, & \alpha^J \pi < |\omega| \leq \pi \end{cases} \tag{7.17}$$

$$H_1^J(\omega) = \begin{cases} H_1\left(\frac{\omega}{\alpha^{J-1}}\right) \prod_{m=0}^{J-2} H_0\left(\frac{\omega}{\alpha^m}\right), & (1-\beta)\alpha^{J-1}\pi \leq |\omega| \leq \alpha^{J-1}\pi \\ 0, & \text{for other } \omega \in [-\pi, \pi] \end{cases} \tag{7.18}$$

The $r$ and Q-factor are related with filter bank parameters $\alpha$ and $\beta$ in the following manner:

$$r = \frac{\beta}{1-\alpha}, \ Q = \frac{2-\beta}{\beta}$$

The Q-factor of the TQWT at some level $j$ can be defined as, $Q = \frac{f_c(j)}{BW(j)}$, where $f_c(j)$ is the center frequency and $BW(j)$ is the bandwidth at level $j$.
Where,

$$f_c(j) = \alpha^j \frac{2-\beta}{4\alpha} f_s, \ j = 1, 2, ..., J \tag{7.19}$$

$$BW(j) = \frac{1}{2}\beta\alpha^{j-1}\pi, \ j = 1, 2, ..., J \tag{7.20}$$

The original signal can be obtained by using synthetic filter bank which has been shown in Fig. 7.8 (c). Perfect reconstruction of the signal from the TQWT coefficients can be performed if the filters $H_0(\omega)$ and $H_1(\omega)$ are chosen such that the spectrum of $s(n)$ will be equal to the spectrum of the reconstructed signal $y(n)$ in Fig. 7.8. The relation between the spectrum of input signal $s(n)$ and spectrums of $y_0(n)$ and $y_1(n)$ in Fig. 7.8 are given by,

$$Y_0(\omega) = \begin{cases} |H_0(\omega)|^2 S(\omega), & |\omega| \leq \alpha\pi \\ 0, & \alpha\pi < |\omega| \leq \pi \end{cases} \tag{7.21}$$

and

$$Y_1(\omega) = \begin{cases} 0, & |\omega| \leq (1-\beta)\pi \\ |H_1(\omega)|^2 S(\omega), & (1-\beta)\pi < |\omega| \leq \pi \end{cases} \tag{7.22}$$

The above Eqs. (7.21) and (7.22) can be combined to obtain the spectrum corresponding to $y(n)$.

$$Y(\omega) = \begin{cases} |H_0(\omega)|^2 S(\omega), & |\omega| \leq (1-\beta)\pi \\ (|H_0(\omega)|^2 + |H_1(\omega)|^2)S(\omega), & (1-\beta)\pi \leq |\omega| \leq \alpha\pi \\ |H_1(\omega)|^2 S(\omega), & \alpha\pi < |\omega| \leq \pi \end{cases} \tag{7.23}$$

From Eq. (7.23), the criteria for the low-pass and high-pass filters for perfect reconstruction can be derived which are given as follows:

$$\begin{cases} |H_0(\omega)| = 1, & |\omega| \leq (1-\beta)\pi \\ H_0(\omega) = 0, & \alpha\pi < |\omega| \leq \pi \end{cases}$$

$$\begin{cases} H_1(\omega) = 0, & |\omega| \leq (1-\beta)\pi \\ |H_1(\omega)| = 1, & \alpha\pi < |\omega| \leq \pi \end{cases}$$

The above criterion are given for the pass-bands of two filters. In transition band $((1-\beta)\pi \leq |\omega| \leq \alpha\pi)$, $Y(\omega) = S(\omega)$ only if $|H_0(\omega)|^2 + |H_1(\omega)|^2 = 1$. Any power complementary function $\phi(\omega)$ with period $2\pi$ can be used to design the transition bands of $H_0(\omega)$ and $H_1(\omega)$. $\phi(\omega)$ will satisfy the condition, $\phi^2(\omega) + \phi^2(\pi - \omega) = 1$. The scaled and translated version of $\phi^2(\omega)$ and $\phi^2(\pi - \omega)$ from the interval $[0, \pi]$ to the interval of transition bands of the low-pass and high-pass filters can be used to design the transition functions of these filters in terms of $\phi(\omega)$. The transition functions of the filters $H_0(\omega)$ and $H_1(\omega)$ are given as follows:

$$H_0(\omega) = \phi\left(\frac{\omega + (\beta - 1)\pi}{\alpha + \beta - 1}\right) \tag{7.24}$$

$$H_1(\omega) = \phi\left(\frac{\alpha\pi - \omega}{\alpha + \beta - 1}\right) \tag{7.25}$$

Here, $\phi(\omega)$ is the frequency response of the Daubechies filter having two vanishing moments. The $\phi(\omega)$ can be expressed as,

$$\phi(\omega) = 0.5[1 + \cos(\omega)]\sqrt{2 - \cos(\omega)}, |\omega| \leq \pi \tag{7.26}$$

The effect of parameter $J$ on TQWT-based decomposition when $Q = 2$, $r = 3$, and $J = 3, 6$ has been shown in Fig. 7.9 (a). The number of sub-bands are 3 and 6 for $J = 3$ and 6, respectively. The frequency resolution of filter bank becomes better with increase in $J$. The effect of parameter $Q$ is shown in Fig. 7.9 (b) where $J = 3$, $r = 3$, and $Q = 2, 4$. The change in $Q$ compresses the filter bank. The effect of parameter $r$ when $Q = 2$, $r = 3$ and 5, and $J = 3$ is shown in Fig. 7.9 (c). The increase in $r$ shifts the filter bank toward higher frequency as the overlap between the filters is increased [91].

   As an example, TQWT decomposition of EEG signal is shown in Fig. 7.10 for $Q = 2$, $r = 3$, and $J = 6$. The EEG signal is downloaded from MATLAB (Espiga3) which has sampling frequency of 200 Hz. The TFR obtained by applying HSA on the decomposed components are shown in Fig. 7.11.

## 7.5   FLEXIBLE ANALYTIC WAVELET TRANSFORM

The flexible analytic wavelet transform (FAWT) is a RDWT where the dilation factor, Q-factor, and the redundancy can be easily specified. The FAWT is also known as an analytic wavelet transform with a flexible time-frequency covering [92]. The FAWT can be implemented with the iterative filter bank approach. As compared to other wavelets transforms, it has one low-pass and two high-pass channels at each level in its iterative filter bank. The two high-pass channels separate positive and negative frequencies and give analytic bases. In this method, fractional sampling is also possible in the low-pass and high-pass channels. The above-mentioned properties give adjustable parameters, namely, redundancy ($r$), dilation factor, or Q-factor. The analysis of transient and oscillatory natures of the signals can also be performed with FAWT method. The frequency-domain–based representation of the filters for FAWT method can be given as,

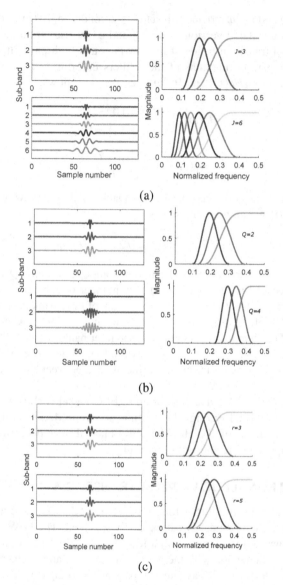

**Figure 7.9** (a) Effect of parameter $J$ on TQWT-based decomposition ($Q = 2$, $r = 3$, and $J = 3$, 6), (b) effect of parameter $Q$ on TQWT-based decomposition ($Q = 2$, 4, $r = 3$, and $J = 3$), and (c) effect of parameter $r$ on TQWT-based decomposition ($Q = 2$, $r = 3$, 5, and $J = 3$).

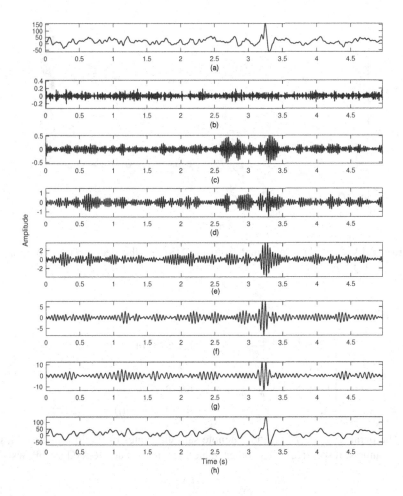

**Figure 7.10** (a) EEG signal in time-domain, (b)–(h) decomposed components obtained using TQWT method.

Low-pass filter:

$$H(\omega) = \begin{cases} (ef)^{1/2}, & \text{for } |\omega| < \omega_p \\ (ef)^{1/2}\theta\left(\dfrac{\omega-\omega_p}{\omega_s-\omega_p}\right), & \text{for } \omega_p \leq \omega \leq \omega_s \\ (ef)^{1/2}\theta\left(\dfrac{\pi-\omega+\omega_p}{\omega_s-\omega_p}\right), & \text{for } -\omega_s \leq \omega \leq -\omega_p \\ 0, & \text{for } |\omega| \geq \omega_s \end{cases} \qquad (7.27)$$

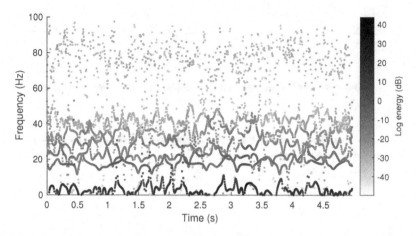

**Figure 7.11** The TFR obtained from the HSA of the decomposed components based on TQWT.

High-pass filter:

$$G(\omega) = \begin{cases} (2uv)^{1/2}\theta\left(\frac{\pi-(\omega+\omega_0)}{\omega_1-\omega_0}\right), & \text{for } \omega_0 \leq \omega < \omega_1 \\ (2uv)^{1/2}, & \text{for } \omega_1 \leq \omega < \omega_2 \\ (2uv)^{1/2}\theta\left(\frac{\omega-\omega_2}{\omega_3-\omega_2}\right), & \text{for } \omega_2 \leq \omega \leq \omega_3 \\ 0, & \text{for } \omega \in [0,\omega_o) \cup (\omega_3,2\pi) \end{cases} \quad (7.28)$$

The parameters $e$, $f$, $u$, and $v$ denote sampling rates of the low-pass and high-pass channels, respectively. The other related parameters are defined as follows:

$$\omega_p = \frac{(1-\beta)\pi+\varepsilon}{e} \quad (7.29)$$

$$\omega_s = \frac{\pi}{f} \quad (7.30)$$

$$\omega_o = \frac{(1-\beta)\pi+\varepsilon}{u} \quad (7.31)$$

$$\omega_1 = \frac{e\pi}{fu} \quad (7.32)$$

$$\omega_2 = \frac{\pi-\varepsilon}{u} \quad (7.33)$$

$$\omega_3 = \frac{\pi+\varepsilon}{u} \quad (7.34)$$

$$\varepsilon \leq \left(\frac{e-f+\beta f}{e+f}\right)\pi \quad (7.35)$$

$$Q = \frac{2-\beta}{\beta} \qquad (7.36)$$

$$1 - \frac{e}{f} \leq \beta \leq \frac{u}{v} \qquad (7.37)$$

In the above-mentioned representations of $H(\omega)$ and $G(\omega)$, $\theta$ must satisfy the following condition:

$$\theta(\omega) = \frac{\sqrt{2-\cos(\omega)}}{2}[1+\cos(\omega)], \quad \text{for } \omega \in [0,\pi] \qquad (7.38)$$

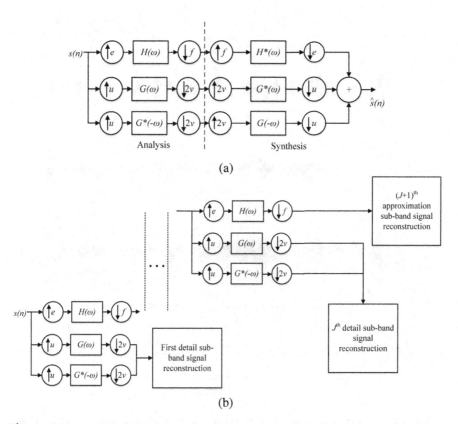

(a)

(b)

**Figure 7.12** (a) Analysis and synthesis processes of signal based on the first-level FAWT and (b) The $J^{\text{th}}$ level FAWT-based decomposition to obtain $J+1$ sub-band signals similar to DWT.

The selection of values $\beta$, $e$, $f$, $u$, and $v$ is required for design of FAWT-based decomposition using filter bank. The Fig. 7.12 (a) shows the analysis and synthesis processes for first-level FAWT process. In the literature, the detail coefficients corresponding to negative and positive parts are combined to form detail coefficients, so that, the FAWT-based method can be applied like conventional DWT method. This modified form of FAWT for a signal $s(n)$ is shown in Fig. 7.12 (b).

Figure 7.13 shows an example of FAWT decomposition of ECG signal. The ECG signal is taken from MIT-BIH arrhythmia database with sampling rate of 360 Hz [3, 4]. The parameters used for designing the FAWT filter bank are $p = 275$, $q = 279, r = 1, s = 20$, and $J = 3$. The TFR obtained by applying HSA on the obtained components from FAWT for the considered ECG signal is shown in Fig. 7.14.

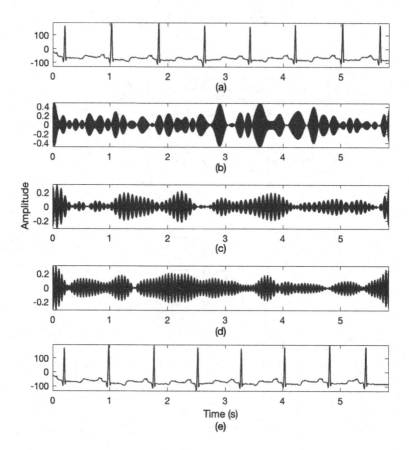

**Figure 7.13** (a) ECG signal and (b)-(e) its decomposed components using FAWT technique.

## 7.6  FBSE-BASED FLEXIBLE ANALYTIC WAVELET TRANSFORM

FBSE-based FAWT (FBSE-FAWT) is improved version of FAWT, which uses FBSE for implementation of FAWT [93]. The TQWT and FAWT form a family of wavelet transform which have the flexibility to tune Q-factor according to application. The

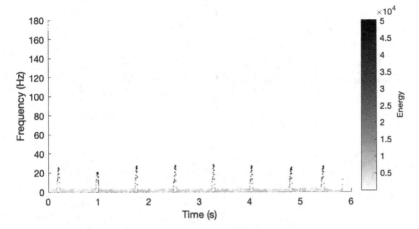

**Figure 7.14** The TFR obtained from FAWT-based decomposed components of considered ECG signal and HSA.

TQWT allows an easy way to select $Q$ and redundancy ($r$). However, for a given $Q$ and $r$, TQWT does not provide flexibility to select the desired dilation factor. FAWT solves this problem by splitting the high-pass channel into negative and positive frequency parts, which allows the selection of an arbitrary sampling rate in a high-frequency channel. So, $Q$, $r$, and dilation factor can be easily selected in FAWT.

In FBSE-FAWT method, the FBSE spectrum is used for the implementation of FAWT instead of Fourier spectrum. The FAWT performs filtering in frequency-domain by multiplying the filter response with the Fourier transform-based spectrum of the signal. In contrast to FAWT, the FBSE-FAWT uses the FBSE spectrum obtained using the relation between order of the coefficients and frequencies as given in Eq. (2.42). The FBSE represents real signal only in terms of positive frequencies; therefore, it makes the implementation process easier, as the high-pass channel will no longer require two separate filters for positive and negative frequency parts (only required filter for positive frequencies). The various advantages of FBSE-based representation over Fourier-based representation are explained in section 2.5. The filter bank for implementation of analysis and synthesis parts of the FBSE-FAWT are shown in Fig. 7.15 (a) and filter bank for iterative implementation of FBSE-FAWT is shown in Fig. 7.15 (b).

The mathematical expressions of $H(\omega)$ and $G(\omega)$ can be given by,

$$H(\omega) = \begin{cases} (ef)^{1/2}, & \omega < \omega_p \\ (ef)^{1/2}\theta\left(\frac{\omega-\omega_p}{\omega_s-\omega_p}\right), & \omega_p \leq \omega \leq \omega_s \\ 0, & \omega \geq \omega_s \end{cases} \tag{7.39}$$

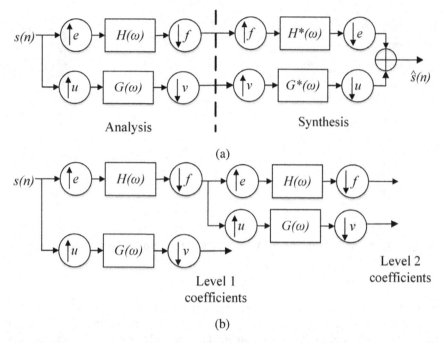

**Figure 7.15** (a) Analysis and synthesis processes of FBSE-FAWT method and (b) iterative implementation of two-level FBSE-FAWT method.

$$
G(\omega) = \begin{cases} (uv)^{1/2}\theta\left(\frac{\pi-(\omega-\omega_0)}{\omega_1-\omega_0}\right), & \omega_0 \le \omega < \omega_1 \\ (uv)^{1/2}, & \omega_1 \le \omega < \omega_2 \\ (uv)^{1/2}\theta\left(\frac{\omega-\omega_2}{\omega_3-\omega_2}\right), & \omega_2 \le \omega \le \omega_3 \\ 0, & \omega \in [0,\omega_o) \cup (\omega_3,2\pi) \end{cases} \tag{7.40}
$$

All parameters in Eqs. (7.39) and (7.40) and in Figs. 7.15 (a) and (b) are same as mentioned in section 7.5.

For analysis, an ECG signal, taken from the MIT-BIH arrhythmia dataset, is considered and decomposed using the FBSE-FAWT technique [3, 4]. The decomposition results are shown in Fig. 7.16. The sampling frequency of the considered signal is 360 Hz. The TFR of the considered ECG signal is obtained using HSA of the decomposed components from FBSE-FAWT technique which has been shown in Fig. 7.17.

## 7.7 DUAL-TREE COMPLEX WAVELET TRANSFORM

Conventional wavelet transform suffers from limitations like, limited directionality, absence of the phase information, poor localization of singularities, and its shift

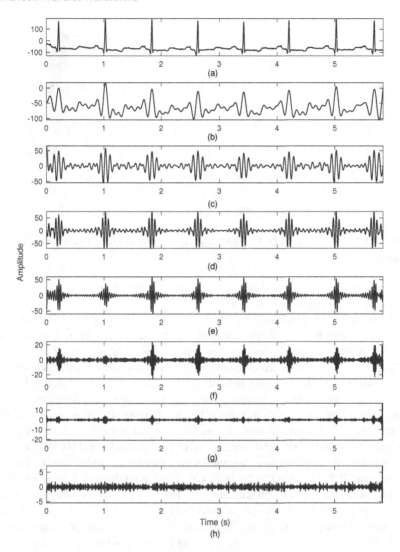

**Figure 7.16** (a) The ECG signal and (b)-(h) its decomposed components obtained using FBSE-FAWT method.

variance nature. These drawbacks of conventional wavelet transform were the motivation for development of complex wavelet transform. The complex wavelet transform uses complex valued scaling and wavelet functions. The complex wavelet can be represented as,

$$\psi_C(t) = \psi_R(t) + j\psi_I(t) \qquad (7.41)$$

where, $\psi_R(t)$ and $\psi_I(t)$ are the real and imaginary parts of the complex wavelet, respectively. Moreover, $\psi_I(t)$ is the Hilbert transform pair of $\psi_R(t)$ in order $\psi_C(t)$ to

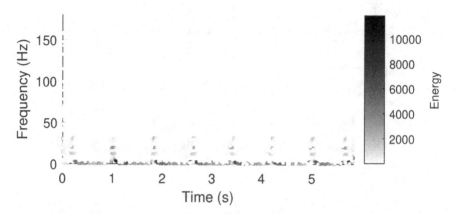

**Figure 7.17** The TFR of ECG signal obtained using HSA of FBSE-FAWT–based components.

be analytic which only have positive frequency components. In a similar manner, the analytic scaling function can be defined.

The complex wavelet transform can be implemented using real DWT followed by two filters which pass only positive and negative frequency parts. This approach will allow positive frequency components of the signal along with small part of negative frequency components also. Another Hilbert transform-based approach for complex wavelet transform computes the Hilbert transform of the input signal. Then on both the input and Hilbert transformed signals, real DWT are performed. Finally, to obtain the complex wavelet transform, the coefficients of each wavelet transform are combined. This approach suffers from the problem of having a long support or poor time-localization. The dual-tree complex wavelet transform (DTCWT) is an other approach for implementing complex wavelet transform which requires two real DWTs in parallel [94]. One DWT provides real part and other DWT provides imaginary part of DTCWT. In implementation, two real DWTs, employ two different sets of filters and each one satisfies perfect reconstruction conditions. Two sets of filters have been designed jointly so that the overall transform is analytic in approximate way. In Fig. 7.18 (a), $h_0(n)$ and $h_1(n)$ represent low-pass and high-pass filter pair for the upper filter bank and $g_0(n)$ and $g_1(n)$ represent low-pass and high-pass filter pair for the lower filter bank. The two real mother wavelets associated with each of the DWTs are denoted as $\psi_h(t)$ and $\psi_g(t)$. In order to satisfy the perfect reconstruction, the filters have been designed such that $\psi(t) = \psi_h(t) + j\psi_g(t)$ is analytic in approximate manner. It should be noted that, these wavelets are designed in such a way that $\psi_g(t)$ is Hilbert transform of $\psi_h(t)$, i.e. [95],

$$\psi_g(t) \approx \mathrm{H}\{\psi_h(t)\} \qquad (7.42)$$

where, $\mathrm{H}\{\cdot\}$ denotes the Hilbert transform operator.

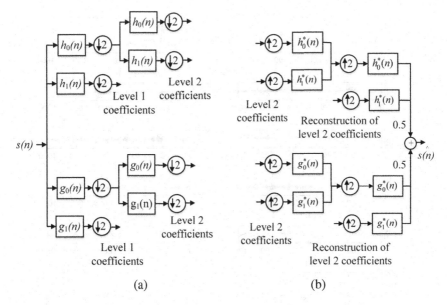

**Figure 7.18** (a) Two-level analysis filter bank for DTCWT and (b) two-level synthesis filter bank for DTCWT.

The above-mentioned criteria will be satisfied if the $g_0(n)$ and $h_0(n)$ are half-sample shift to each other [96, 97], which can be expressed as,

$$g_0(n) \approx h_0(n-0.5)$$
$$G_0(e^{j\omega}) = e^{-j0.5\omega} H_0(e^{j\omega}) \tag{7.43}$$

From Eq. (7.43), half sample delay can be written separately for magnitude and phase functions as follows:

$$|G_0(e^{j\omega})| = |H_0(e^{j\omega})| \tag{7.44}$$

$$\angle G_0(e^{j\omega}) = \angle H_0(e^{j\omega}) - 0.5\omega \tag{7.45}$$

Filters $g_0(n)$ and $h_0(n)$ can be designed in different ways so that the magnitude and phase criteria as given in Eqs. (7.44) and (7.45) will be satisfied. To obtain a linear-phase biorthogonal filter, $h_0(n)$ and $g_0(n)$ can be chosen as a symmetric odd-length (type-I) finite impulse response (FIR) and symmetric even-length (type-II) FIR filter which follow the phase criterion exactly. The coefficients of the filter can be chosen by an iterative error minimization strategy to approximately fulfil the magnitude criterion. The analysis and synthesis parts for DTCWT are shown in Figs. 7.18 (a) and (b), which consist of tree-A (shown in upper part of the Fig. 7.18) and tree-B (shown in lower part of the Fig. 7.18) corresponding to real and imaginary parts of the coefficients.

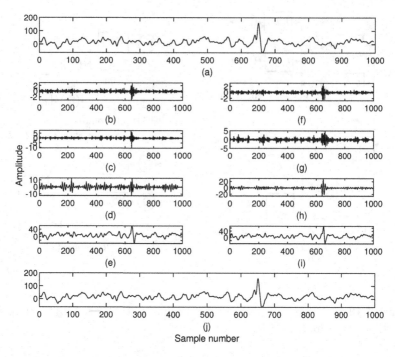

**Figure 7.19** Example for DTCWT decomposition. (a) EEG signal, (b)–(d) $D1$, $D2$, and $D3$ are detail sub-band signals of tree-A for first, second, and third levels, respectively, (e) $A3$ is approximation sub-band signal corresponding to tree-A, (f)–(h) $D1$, $D2$, and $D3$ denote detail sub-band signals of tree-B for first, second, and third levels, respectively, (i) $A3$ denotes approximation sub-band signal corresponding to tree-B, and (j) reconstructed EEG signal.

The DTCWT has the useful properties like as, high directionality, shift invariance, perfect reconstruction, and high computational efficiency. The example of decomposition of EEG signal using DTCWT is shown in Fig. 7.19. Where $D1$, $D2$, $D3$, and $A3$ denote decomposed first, second, and third detail and approximation sub-band signals of EEG signal corresponding to the first, second, and third detail and approximation coefficients, respectively. The EEG signal (Espiga3) with sampling frequency of 200 Hz is obtained from MATLAB.

It should be noted that the DTCWT method is not a critically sampled transform. On the other hand, it is twice expensive than 1D case as the total output data rate is exactly two times of the input data rate. Filters are real and no complex conversion is required for them.

The inverse of DTCWT can be carried out like as the forward transform. In order to invert the transform, the inverse of each of these two real DWTs, have been used in order to obtain the final output. These two real signals have been averaged out in order to get final output. It should be noted that the original signal can be obtained for either real part or imaginary part separately, for real signal.

## PROBLEMS

Q 7.1 A general model of a synthetic signal is given as,

$$f(t) = \cos\left(\omega_0 t + \frac{1}{2}\beta t^2\right)$$

Consider sampling frequency $f_s = 1000$ Hz. Vary the signal parameters $\omega_0$, and $\beta$ so that the signal will lie in the frequency range of 0-400 Hz. Determine the sub-band signal corresponding to 125-375 Hz based on WPT.

Q 7.2 Take a signal,

$$f(t) = \cos(\omega_0 t) + \cos(10\omega_0 t)$$

Determine suitable $Q$ for TQWT-based analysis and obtain corresponding TFR by HSA technique. Consider suitable sampling frequency for simulation.

Q 7.3 Consider any speech signal (you may use mtlb.mat, available in MATLAB). Decompose using TQWT with suitable values of $Q$, $r$, and $J$ parameters and show the sub-band signals in time-domain and frequency-domain. Sketch their distribution of energy with respect to sub-bands. Check energy preservation in TQWT domain.

Q 7.4 Take any speech signal and decompose it using FAWT (4-level). Sketch the sub-band signals and corresponding spectrums. Show that the signal can be obtained back from sub-band signals. Determine the MSE between the original and reconstructed signals. Based on the trial and error method, choose the set of FAWT parameters which is suitable for analyzing the above-mentioned signal.

Q 7.5 Consider multi-component signal,

$$x(t) = \cos(2\pi 100t + 20\pi t^2) + \cos(40\pi t)$$

Decompose the $x(t)$ using both FAWT and TQWT methods. Show the TFRs based on HSA of the above-mentioned signal obtained using FAWT and TQWT. Based on trial and error method, choose the best FAWT and TQWT parameters for TFR. Explain the behavior of TFR for both cases. Consider suitable sampling frequency for simulation.

Q 7.6 Take a bat signal (you may use the bat signal available MATLAB using the command "load batsignal") and obtain TFR using SWT. Compare the TFR based on SWT with spectrogram and scalogram. Show which method is able to identify the chirp components present in the bat signal more accurately.

Q 7.7 For the signal $x(n)$ defined as,

$$x(n) = \begin{cases} \cos(20\pi\frac{n}{f_s}), & \text{if } 0 \leq n < 500 \\ \cos(20\pi\frac{n}{f_s}) + \sin(230\pi\frac{n}{f_s}), & \text{if } 500 \leq n < 1000 \\ \cos(20\pi\frac{n}{f_s}), & \text{if } 1000 \leq n < 1500 \\ \cos(100\pi\frac{n}{f_s}), & \text{if } 1500 \leq n < 2000 \\ 0, & \text{otherwise} \end{cases}$$

where, sampling frequency $f_s = 1$ kHz. Compute the TFR of the signal $x(n)$ using FBSE-FAWT method. Also discuss the differences between TFR based on FBSE-FAWT and scalogram.

Q 7.8 Perform signal separation of the following multi-component signals using DTCWT method:

(a) $x(n) = \cos\left[40\pi\frac{n}{f_s} + 10\cos(2\pi\frac{n}{f_s})\right] + \left[1 + \sin(2\pi\frac{n}{f_s})\right]\cos(100\pi\frac{n}{f_s})$

(b) $x(n) = \cos(20\pi\frac{n}{f_s} + 10(\frac{n}{f_s})^2) + \cos(100\pi\frac{n}{f_s} + 10(\frac{n}{f_s})^2)$

where, $n = 1, 2, 3, \ldots, 1000$ and sampling frequency $f_s$ is 1 kHz. Check whether DTCWT can separate the signal components or not.

Q 7.9 Consider a multi-component signal $x(t)$, $0 \leq t \leq 3$ defined as,

$$x(t) = \begin{cases} \cos[2\pi(10 + 60t^2)t], & \text{if } 0 \leq t < 1 \\ \cos[2\pi(10 + 60t^2)t] + \cos[2\pi(20 + 105t^2)t], & \text{if } 1 \leq t < 2 \\ \cos[2\pi(10 + 60t^2)t], & \text{if } 2 \leq t \leq 3 \end{cases}$$

Sample the signal with suitable sampling rate and add 20 dB additive white Gaussian noise (AWGN). Perform the threshold-based denoising of this noisy signal using WPT, DWT, and TQWT. Also sketch the plot of MSE versus threshold value for the reconstructed signal with respect to original signal.

Q 7.10 Consider the signal $x(n)$ expressed below.

$$x(n) = k(n)[5\cos(2\pi 20n/256) + 5\cos(2\pi 60n/256)]$$

where,

$$k(n) = \begin{cases} 1 - (n - 128)^2/400, & \text{if } 108 \leq n \leq 148 \\ 0, & \text{otherwise} \end{cases}$$

Based on the distribution of the energy in frequency-domain of the signal $x(n)$, determine the wavelet packets for obtaining the signal components. Based on these frequency ranges, separate the components by WPT. Consider the sampling rate of $x(n)$ as 1.

Q 7.11 Take ECG signal (from any publicly available database) and decompose using FAWT. Modify the FAWT process to obtain DWT like decomposition in following way: Add two detail coefficients to get single detail coefficients at each level. Obtain TFRs using HSA from the 4-level decomposed components based on FAWT and modified FAWT and compare both the TFRs.

Q 7.12 Study the effect of scaling operation $x(n) \rightarrow x(2n)$ on WPT, TQWT, FAWT, and DTCWT based decomposed components and their spectrums for any considered arbitrary signal.

# 8 Adaptive Time-Frequency Transforms

*"Unless you try to do something beyond what you have already mastered, you will never grow."* –Ralph Waldo Emerson

## 8.1 HILBERT-HUANG TRANSFORM

The Hilbert-Huang transform (HHT) is a two-step process for the analysis of nonlinear and non-stationary signals [98]. The first step is empirical mode decomposition (EMD) which decomposes any signal into a set of band-limited AFM signals known as intrinsic mode functions (IMFs) where highest frequency component is represented by the first IMF and lowest frequency component is represented by the last IMF. The second part of HHT is Hilbert transform which converts obtained IMFs from EMD into analytic IMFs that can be used to determine AE and IF functions of these IMFs. The arrangement of AE and IF functions of these IMFs provides a TFR which is termed as HHT.

### 8.1.1 EMPIRICAL MODE DECOMPOSITION

EMD method is an adaptive and data dependent technique which does not require conditions about stationarity and linearity for signal analysis. This method decomposes any time-domain signal into a finite number of AFM components which represent the basis functions for EMD-based decomposition. The IMFs which are obtained using sifting process should satisfy the following conditions:

Condition 1: In the complete data set, the number of extrema and zero-crossings must be either equal or differ at most by one.

Condition 2: At any point, the mean value of the envelopes obtained from maxima and minima should be zero.

Based on the EMD method, the non-stationary signal $x(t)$ can be represented as a sum of IMFs and residue component as,

$$x(t) = \sum_{j=1}^{J} \text{IMF}_j(t) + r_J(t) \tag{8.1}$$

In this expression, the $\text{IMF}_j(t)$ represents the $j^{\text{th}}$ IMF and $r_J(t)$ denotes the residue part of this decomposition.

DOI: 10.1201/9781003367987-8

The EMD method can be summarised in terms of required steps as follows:

Step 1: Obtain all extrema of $x(t)$.

Step 2: Obtain the upper envelope $E_{max}(t)$ and lower envelope $E_{min}(t)$ by interpolating the maxima and minima based on cubic spline interpolation method, respectively.

Step 3: Compute the average of $E_{max}(t)$ and $E_{min}(t)$ using the following equation:

$$A(t) = \frac{E_{min}(t) + E_{max}(t)}{2} \tag{8.2}$$

Step 4: Obtain the residue as,

$$D(t) = x(t) - A(t) \tag{8.3}$$

Step 5: Check whether $D(t)$ is satisfying the above-mentioned conditions of IMF,

1. If yes, then $D(t)$ is an IMF of the signal $x(t)$ and for computation of the next IMF, $x(t)$ is replaced with the residue signal $x(t) - D(t)$ and the procedure from the step 1 is repeated.

2. If not, then replace $x(t)$ with $D(t)$ and repeat the process from step 1. To terminate the iteration process, the standard deviation $\sigma_k$ of $D(t)$ of two consecutive iterations can lie in the range 0.2-0.3, where

$$\sigma_k = \sum_{t=0}^{B} \left[ \frac{|D_{k-1}(t) - D_k(t)|^2}{D_{k-1}^2(t)} \right] \tag{8.4}$$

$B$ is the duration of the signal, and $D_k(t)$ represents $D(t)$ after $k^{th}$ iteration.

In step 5, if the residue signal $x(t) - D(t)$ is a monotonic function or further IMFs can not be derived; EMD process is terminated.

The IMFs, namely $IMF_1(t)$, $IMF_2(t)$, ..., $IMF_N(t)$, include different frequency bands from high-frequency to low-frequency. The frequency components contained in each frequency band are different and they change with the input signal due to the adaptive nature of the decomposition. The $r_J(t)$ represents residue of the signal.

For example, Fig. 8.1 shows the synthetic signal $x(n)$ having sampling frequency, $f_s = 1000$ Hz and the IMFs obtained using EMD. The mathematical expression of $x(n)$ is given as,

$$x(n) = \cos\left[2\pi\left(30 + 45\frac{n}{f_s}\right)\frac{n}{f_s}\right] + 0.85\cos\left[2\pi\left(100 + \frac{115}{3}\frac{n^2}{f_s^2}\right)\frac{n}{f_s}\right]$$

where, $n = 0, 1, 2, \ldots, 1000$.

The analytic IMFs are obtained using the Hilbert transform. It should be noted that each IMF is a narrow-band component and can be considered as a mono-component signal. The analytic form of each IMF using Hilbert transform can be computed using Eq. (3.64).

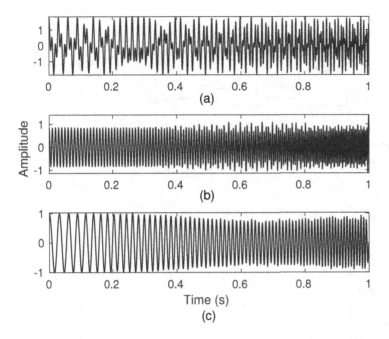

**Figure 8.1**    (a) Time-domain representation of signal $x(n)$, (b)–(c) IMFs obtained from EMD of the signal.

The AE and instantaneous phase parameters are computed using the expressions shown in Eq. (3.61). The $i^{\text{th}}$ IMF can be expressed in terms of AFM model as,

$$\text{IMF}_i(t) = \Re\{a_i(t)e^{j\int_0^t \omega_i(\tau)d\tau}\} \tag{8.5}$$

Further, the signal $x(t)$ can be represented as sum of IMFs as,

$$x(t) = \Re\left\{\sum_{i=1}^{J} a_i(t)e^{j\int_0^t \omega_i(\tau)d\tau}\right\} \tag{8.6}$$

Here, residue $r_J(t)$ is neglected. The residue represents a mean trend.

The TFR of $i^{\text{th}}$ IMF, $\text{HHT}_i(t, \omega)$ can be obtained using HSA, represented mathematically by Eq. (3.74). In the end, the HSA of $x(t)$, $\text{HHT}(t, \omega)$, can represent TFR as a sum of TFRs of IMFs, $\text{HHT}_i(t, \omega)$ as,

$$\text{HHT}(t, \omega) = \sum_{i=1}^{J} \text{HHT}_i(t, \omega) \tag{8.7}$$

Figure 8.2 shows the block diagram of the HHT process. It should be noted that HHT distributes the energy of the signal in time-frequency plane. Figure 8.3 shows the HHT-based TFR of the above-mentioned synthetic signal $x(n)$.

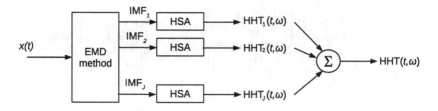

**Figure 8.2** Block diagram of HHT process to obtain TFR. Here, HSA represents the extraction of AE and IF functions of the IMFs using analytic signal representation.

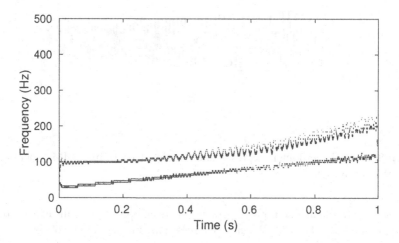

**Figure 8.3** TFR obtained from the IMFs extracted using EMD of the signal $x(n)$.

The EMD method suffers from the following problems such as: mode mixing problem, lack of mathematical theory, boundary distortion problem, and inability in separation of closely spaced frequency components [99].

## 8.2    ENSEMBLE EMPIRICAL MODE DECOMPOSITION

EMD method for signal decomposition has some issues like mode mixing and small perturbation in input signal which can lead to a completely different set of IMFs. In order to produce stable IMFs and mode mixing free IMFs, the ensemble EMD (EEMD) method for signal analysis is proposed. In EEMD method, the addition of white noise to signal is required. The following steps are required in order to perform signal decomposition based on EEMD method [100, 101]:

Step 1: Perform addition of white noise to the signal.

Step 2: Perform the decomposition of signal with added white noise into a set of IMFs.

Step 3: Repeat steps 1 and 2 again and again but every time different white noise needs to be considered.

Step 4: Obtain the ensemble means of the obtained IMFs at each additive noise case.

It should be noted that in the EEMD method during computation of final IMFs from the mean of corresponding IMFs, the noise signals cancel each other. In this way the obtained IMFs are free from mode-mixing effect.
The above-mentioned process for EEMD method can be expressed as follows:
Suppose we perform ensemble of $N$ white noise, $w_p(n)$ at a particular signal to noise ratio (SNR). The ensemble process for EEMD method is explained below.
The noisy signal $x_{dp}(n)$ for $N$ cases is expressed as follows:

$$x_{dp}(n) = x_d(n) + w_p(n), \quad p = 1, 2, \ldots, N \tag{8.8}$$

Now, EMD method is applied to each of these signals to obtain their corresponding IMFs.

$$x_{dp}(n) = \sum_{i=1}^{J} \text{IMF}_{pi}(n) + r_{pJ}(n), \quad p = 1, 2, \ldots, N \tag{8.9}$$

The final IMFs (FIMFs) obtained using EEMD method are given by,

$$\text{FIMF}_i(n) = \frac{1}{N} \sum_{p=1}^{N} \text{IMF}_{pi}(n), \quad i = 1, 2, \ldots, J \tag{8.10}$$

Figure 8.4 shows the obtained IMFs for seizure-free EEG signal of sampling frequency 173.61 Hz taken from Bonn EEG database where $N = 10$ has been considered.
The obtained FIMFs can be converted into analytic FIMFs using Hilbert transform method as discussed earlier in HHT method and in similar way TFR can be obtained. Figure 8.5 shows the TFR obtained using HHT method and EEMD-based method. The major limitation of EEMD method is that it suffers from huge computational complexity problem and difficult to implement it for real-time applications.

## 8.3   VARIATIONAL MODE DECOMPOSITION

The variational mode decomposition (VMD) method overcomes the limitations of EMD method like as empirical nature of the decomposition, mode mixing effect, and cope with noise [102]. The VMD method uses the concept of Wiener filtering, and due to this reason, it is robust to noise. The VMD method decomposes a real-valued

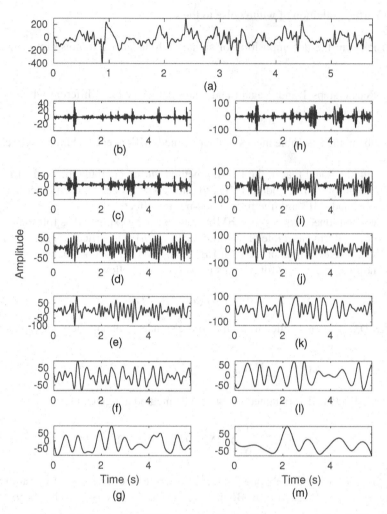

**Figure 8.4** (a) Seizure-free EEG signal, (b)–(g) its corresponding IMFs obtained from EEMD method, and (h)-(m) its corresponding IMFs obtained from EMD method.

signal $x(t)$ into a set of narrowband components (NBCs) $x_l(t)$ and each NBC is centered around its corresponding center frequency. The VMD method first formulates the constrained variational optimization problem in order to estimate the bandwidth of NBCs. The formulation of constrained variational optimization problem can be explained using the following steps:

Step 1: Apply the Hilbert transform operation on each component $x_l(t)$ in order to obtain its single-sided frequency spectrum.

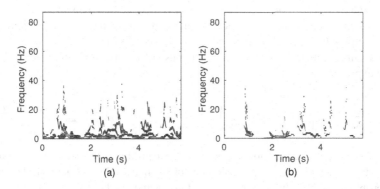

**Figure 8.5** (a) The TFR obtained from HSA of the IMFs obtained from EEMD of the seizure-free EEG signal and (b) the TFR of the same signal obtained using HHT.

Step 2: The modulation property is applied for each component in order to shift its frequency spectrum.

Step 3: In the end, the bandwidth estimation of each component is performed based on the $H^1$ Gaussian smoothness of the demodulated signal.

Mathematically, the constrained variational optimization problem based on the above-mentioned steps is given by,

$$\min_{\{x_l\},\{\omega_l\}} \sum_l \left\| \partial_t \left[ \left( \delta(t) + \frac{j}{\pi t} \right) * x_l(t) \right] e^{-j\omega_l t} \right\|_2^2 \qquad (8.11)$$

such that,

$$\sum_l x_l(t) = x(t) \qquad (8.12)$$

The notation $*$ denotes the convolution process. The set of decomposed components are represented by $\{x_l\} = \{x_1, x_2, \ldots, x_L\}$ and corresponding center frequencies are given by $\{\omega_l\} = \{\omega_1, \omega_2, \ldots, \omega_L\}$, respectively. $\delta(t)$ denotes the Dirac delta function and total number of NBCs obtained from VMD is denoted by $L$. It should be noted that the above-mentioned constrained problem can be transformed into unconstrained optimization problem based on Lagrange multiplier denoted by $\lambda$ and penalty factor is represented by $\alpha$.

The mathematical formulation of unconstrained optimization problem can be given as,

$$\mathscr{L}(\{x_l\}, \{\omega_l\}, \lambda) = \alpha \sum_l \left\| \partial_t \left[ \left( \delta(t) + \frac{j}{\pi t} \right) * x_l(t) \right] e^{-j\omega_l t} \right\|_2^2$$
$$+ \left\| x(t) - \sum_l x_l(t) \right\|_2^2 + \left\langle \lambda(t), x(t) - \sum_l x_l(t) \right\rangle \qquad (8.13)$$

The alternate direction method of multipliers (ADMM) has been applied to convert the above unconstrained optimization problem into sub-optimization problem in order to compute the updated NBCs and associated center frequencies.

The first sub-optimization problem which is used for computing the updated NBCs can be expressed as,

$$x_l^{n+1} = \underset{x_l \in X}{\operatorname{argmin}} \left\{ \alpha \left\| \partial_t \left[ \left( \delta(t) + \frac{j}{\pi t} \right) * x_l(t) \right] e^{-j\omega_l t} \right\|_2^2 + \left\| x(t) - \sum_i x_i(t) + \frac{\lambda(t)}{2} \right\|_2^2 \right\}$$

(8.14)

where $X$ represents the functional space which means that in this space the signal $x(t)$ and signal modes $x_l(t)$ are integrable and square integrable upto second derivative. This problem can be solved in frequency-domain and can be expressed as,

$$\hat{x}_l^{n+1}(\omega) = \frac{\hat{X}(\omega) + \frac{\hat{\lambda}(\omega)}{2} - \sum_{i \neq l} \hat{x}_i(\omega)}{1 + 2\alpha(\omega - \omega_l)^2}$$

(8.15)

Here, $\hat{X}(\omega)$, $\hat{x}_i(\omega)$, $\hat{\lambda}(\omega)$, and $\hat{x}_l^{n+1}$ denote the Fourier transforms of $x(t)$, $x_i(t)$, $\lambda(t)$, and $x_l^{n+1}(t)$, respectively. It should be noted that the above-mentioned expression represents the Wiener filter structure. In the end, the updated components are obtained in time-domain using real part of the inverse Fourier transform of the mentioned expression.

The second sub-optimization problem includes only the bandwidth term. The second sub-optimization problem can be expressed mathematically as,

$$\omega_l^{n+1} = \underset{\omega_l}{\operatorname{argmin}} \left\{ \left\| \partial_t \left[ \left( \delta(t) + \frac{j}{\pi t} \right) * x_l(t) \right] e^{-j\omega_l t} \right\|_2^2 \right\}$$

(8.16)

The center frequencies of the updated components are computed using above expression as,

$$\omega_l^{n+1} = \frac{\int_0^\infty \omega |\hat{x}_l(\omega)|^2 d\omega}{\int_0^\infty |\hat{x}_l(\omega)|^2 d\omega}$$

(8.17)

The VMD method requires selection of number of iterations in order to identify the updated NBCs and their corresponding center frequencies. The estimation of center frequencies can be performed based on the average of center frequencies obtained from all the iterations of VMD method.

The VMD method requires selection of input parameters, namely, penalty factor ($\alpha$), time step of the dual ascent ($z$), number of extracted NBCs, tolerance for convergence criteria (tol), number of extracted DC components (DC), and initial frequencies for the extracted components ($\omega_{int}$).

The VMD-based method for the decomposition of speech signal taken from CMU ARCTIC database having sampling frequency of 16 kHz is shown in the Fig. 8.6. The parameters of VMD for simulation are $\alpha = 1000$, number of extracted NBCs = 5, tol = $5e - 6$, and $\omega_{int}$ are set as frequencies corresponding to the peaks in the magnitude Fourier spectrum of the signal. In similar way to EMD and EEMD, HSA can be used on VMD-based decomposed components in order to obtain TFD.

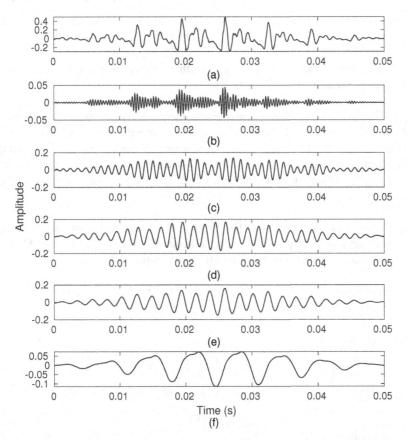

**Figure 8.6** (a) A time-domain voiced speech signal multiplied with Hamming window of signal length, (b)–(f) NBCs of the signal obtained from VMD method.

## 8.4 EMPIRICAL WAVELET TRANSFORM

Previous methods for signal decomposition like as EMD can decompose a signal into a set of narrowband modes. The EMD method is useful for many applications besides lacking mathematical theory. The EMD method suffers from mainly three limitations namely, boundary effect, mode-mixing, and stop criteria. Recently, the method has been proposed for signal analysis based on the adaptive wavelet design for the spectrum. Here, different modes of the signal can be obtained from the designed suitable wavelet filter bank. This formulation results a new wavelet transform and termed as empirical wavelet transform (EWT) [103]. The EWT is an adaptive signal decomposition technique for analysis of non-stationary signals. The inherent principle of EWT method is based on the formation of adaptive wavelet-based filters. These wavelet-based filters contain the support in the spectrum information location corresponding to the analyzed signal. It should be noted that the obtained sub-band signals based on EWT method have specific center frequencies and compact frequency supports. The EWT method requires the following steps:

Step 1: The fast Fourier transform (FFT) method is used to obtain the spectrum in the frequency range $[0, \pi]$.

Step 2: The frequency spectrum is segmented into $N$ number of contiguous segments based on EWT boundary detection method so that optimal set of boundary frequencies can be obtained. The scale-space–based boundary detection method is used for the segmentation purpose. The first and last boundary frequencies are set to 0 and $\pi$, respectively. For other boundary frequency detection, the boundary detection method is employed.

Step 3: The design of empirical scaling and wavelet functions is carried out for each segment as set a of band-pass filters. The concept of Littlewood-Paley and Mayer's wavelets are used for the construction of the wavelet-based filters.

The mathematical expressions for empirical scaling function $\Lambda_i(\omega)$ and wavelet function $\theta_i(\omega)$ are expressed as follows:

Scaling function:

$$\Lambda_i(\omega) = \begin{cases} 1, & \text{if } |\omega| \leq (1-\zeta)\omega_i \\ \cos\left[\dfrac{\pi\eta(\zeta,\omega_i)}{2}\right], & \text{if } (1-\zeta)\omega_i \leq |\omega| \leq (1+\zeta)\omega_i \\ 0, & \text{otherwise} \end{cases} \tag{8.18}$$

Wavelet function:

$$\theta_i(\omega) = \begin{cases} 1, & \text{if } (1+\zeta)\omega_i \leq |\omega| \leq (1-\zeta)\omega_{i+1} \\ \cos\left[\dfrac{\pi\eta(\zeta,\omega_{i+1})}{2}\right], & \text{if } (1-\zeta)\omega_{i+1} \leq |\omega| \leq (1+\zeta)\omega_{i+1} \\ \sin\left[\dfrac{\pi\eta(\zeta,\omega_i)}{2}\right], & \text{if } (1-\zeta)\omega_i \leq |\omega| \leq (1+\zeta)\omega_i \\ 0, & \text{otherwise} \end{cases} \tag{8.19}$$

The function $\eta(\zeta,\omega_i)$ can be expressed as,

$$\eta(\zeta,\omega_i) = \psi\left[\frac{|\omega|-(1-\zeta)\omega_i}{2\zeta\omega_i}\right] \tag{8.20}$$

Here, $\psi(z)$ is an arbitrary function which can be derived as,

$$\psi(z) = \begin{cases} 0, & \text{if } z \leq 0 \\ \psi(z) + \psi(1-z) = 1, & \forall z \in (0,1) \\ 1, & \text{if } z \geq 1 \end{cases} \tag{8.21}$$

The parameter $\zeta$ in the above expressions makes sure that empirical wavelets and scaling function form a tight frame in $L_2(\mathfrak{R})$ and the condition for tight frame can be expressed as,

$$\zeta < \min_i \left( \frac{\omega_{i+1} - \omega_i}{\omega_{i+1} + \omega_i} \right) \tag{8.22}$$

The detail and approximation coefficients can be computed by performing the inner product of the analyzed signal $y(t)$ with wavelets and scaling function, which are represented mathematically as follows:

$$V_{y,\theta}(i,t) = \int y(\tau)\theta_i^*(\tau - t)d\tau \tag{8.23}$$

$$V_{y,\Lambda}(0,t) = \int y(\tau)\Lambda_1^*(\tau - t)d\tau \tag{8.24}$$

Here, $V_{y,\theta}(i,t)$ represents the detail coefficients of $i^{\text{th}}$ level and $V_{y,\Lambda}(0,t)$ denotes the approximation coefficients.

The reconstructed sub-band signals can be expressed as follows:

$$y_0(t) = V_{y,\Lambda}(0,t) * \Lambda_1(t) \tag{8.25}$$

$$y_i(t) = V_{y,\theta}(i,t) * \theta_i(t) \tag{8.26}$$

Where, $y_0(t)$ is the approximation sub-band signal and $y_i(t)$ represents the detail sub-band signal of $i^{\text{th}}$ level. The asterisk ($*$) symbol represents the convolution operation here.

After obtaining components, the HSA can be applied to obtain TFA like in HHT method. A multicomponent synthetic signal $y(n)$, the decomposed components (IMFs), and the TFR obtained using HSA of the IMFs are shown in Figs. 8.7 (a), (b), and (c), respectively. Figure 8.7 (d) shows the TFR of $y(n)$ using EWT and HSA methods. The mathematical model of signal $y(n)$ is given by the following equation:

$$y(n) = \sin\left[100\pi\frac{n}{f_s}\right] + 0.7\cos\left[2\pi\left(100 + \frac{65}{2}\frac{n}{f_s}\right)\frac{n}{f_s}\right]$$

where $n = 1,2,\ldots,1000$ and $f_s = 1000$ samples/s

## 8.5   FBSE-BASED EMPIRICAL WAVELET TRANSFORM

The FBSE-based EWT (FBSE-EWT) method is an improved version of EWT method for non-stationary signal analysis [104]. Here, instead of FFT, the order-zero FBSE method is used for obtaining spectrum and normalized Hilbert transform (NHT) is used for obtaining TFR from the decomposed components. The FBSE spectrum has various advantages over Fourier spectrum which have been explained in section 2.5. The analysis and synthesis equations for order-zero FBSE of discrete-time signal $x(n)$ is represented in Table 2.3. The order ($k$) of the FBSE coefficients

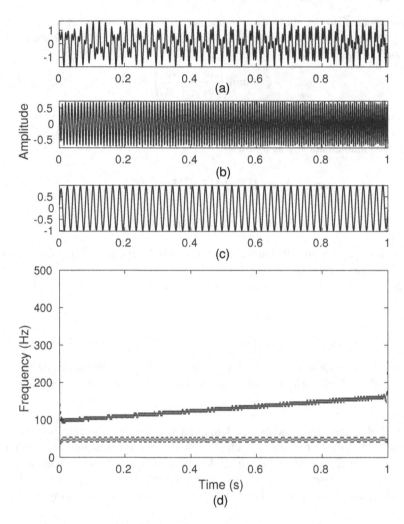

**Figure 8.7**  (a) Signal $x_1(n)$, (b)–(c) IMFs of signal obtained using EWT method, and (d) TFR based on HSA of the decomposed IMFs obtained from EWT.

and corresponding frequency $f_k$ in Hz mapping is given in Eqs. (2.39) and (2.41). In order to cover the entire frequency components of the signal, the FBSE order $L$ should be equal to the length of the signal, i.e., number of samples in the signal.

The FBSE-EWT uses FBSE spectrum instead of FFT spectrum for boundary detection using scale-space–based approach and Otsu's method. Then, using the relation between FBSE order and frequency, the detected FBSE boundaries are mapped into Fourier frequencies. Next to this, the filter bank design and signal component extraction is exactly same as in case of EWT technique. Now, in order to obtain TFR, the NHT is used. The NHT is explained in following subsection.

## 8.5.1   NORMALIZED HILBERT TRANSFORM

The NHT has been developed in order to get rid of Bedrosian condition. In this method, AFM signals are empirically decomposed into AM and phase (FM) parts. The empirical AFM decomposition process is summarized as follows:
This method requires identification of all local maxima points for the absolute value of AFM signal $x(t)$ and connect them with a cubic spline. The spline curve $e_1(t)$ is termed as the envelope of the signal and this obtained envelope is used to normalize the signal $x(t)$. This normalized signal can be expressed as,

$$x_1(t) = \frac{x(t)}{e_1(t)}$$

After normalization, the signal $x_1(t)$ should satisfy the condition $|x_1(t)| \leq 1$. Otherwise envelope of $x_1(t)$ is computed and normalization process is repeated again and expressed as,

$$x_2(t) = \frac{x_1(t)}{e_2(t)}$$

After $n^{th}$ iteration, suppose normalized signal $x_n(t)$ satisfies the condition $|x_n(t)| \leq 1$. After completing the normalization process, the $x_n(t)$ is the empirically derived FM part, which can be represented as,

$$F(t) = x_n(t) = \cos[\phi(t)]$$

The AM part is determined by,

$$A(t) = e_1(t)e_2(t)\ldots e_n(t)$$

Finally, the AFM signal is given by,

$$x(t) = A(t)F(t) = A(t)\cos[\phi(t)]$$

The IF can be computed from FM part based on the analytic signal representation. After obtaining AE and IF functions, the TFR is obtained as,

$$TF(t,f) = \sum_{i=1}^{N} A_i(t)\delta[f - f_i(t)] \tag{8.27}$$

where $A_i(t)$ and $f_i(t)$ denote the AE and IF functions of $i^{th}$ oscillatory level.
Figure 8.8 shows the signal and obtained IMFs using FBSE-EWT method for a synthetic signal $x(n)$. The mathematical model of $x(n)$ is shown in the following equation:

$$x(n) = 1.8\left[1 + 0.8\cos\left(6\pi\frac{n}{f_s}\right)\right]\cos\left(80\pi\frac{n}{f_s}\right)$$
$$+ 1.5\left[1 + 0.6\cos\left(12\pi\frac{n}{f_s}\right)\right]\cos\left(180\pi\frac{n}{f_s}\right),$$

where $n = 1,2,\ldots,1000$ and sampling frequency $f_s = 1000$ samples/s. Figure 8.9 presents the FBSE-EWT and HSA based TFR of signal $x(n)$.

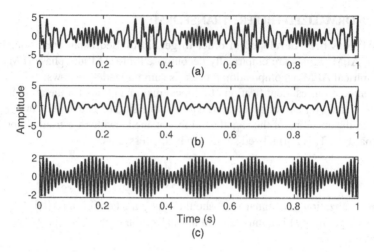

**Figure 8.8**    (a) Signal $x(n)$, and (b)–(c) decomposed components of signal obtained from FBSE-EWT method.

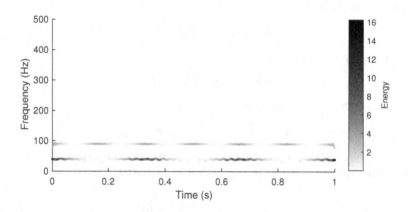

**Figure 8.9**    TFR of signal $x(n)$ obtained using FBSE-EWT method and HSA.

## 8.6   FOURIER DECOMPOSITION METHOD

The Fourier decomposition method (FDM) is a technique to analyze nonlinear and non-stationary signals by using the concept of Fourier representation [105]. It decomposes a signal into a set of Fourier intrinsic band functions (FIBFs). This method presents a generalized Fourier expansion of the signal, i.e., representation of a signal into AFM signals by using the Fourier method itself. The FDM method together with the Hilbert transform provides TFD of a signal which helps localizing it in both time and frequency planes.

## 8.6.1   FOURIER INTRINSIC BAND FUNCTION

Consider a signal $x(t)$, which is defined in interval $[a,b]$ and follows the Dirichlet's conditions. In order to be $x_i(t)$ with $1 \leq i \leq M$ as FIBFs of $x(t)$, the following conditions need to be satisfied:

Condition 1: $x(t) = \sum_{i=1}^{M} x_i(t) + a_0$, where $a_0$ denotes the mean value of $x(t)$.

Condition 2: $\int_a^b x_i(t)dt = 0$, which means FIBFs are zero-mean functions.

Condition 3: $\int_a^b x_i(t)x_j(t)dt = 0$ for $i \neq j$ which means the FIBFs follow orthogonality property.

Condition 4: $x_i(t) + j\mathrm{H}\{x_i(t)\} = a_i(t)e^{j\phi_i(t)}$, where IF $\omega_i(t) = \frac{d}{dt}\phi_i(t) \geq 0$, $\forall\, t$ and AE $a_i(t) \geq 0$, $\forall\, t$, $\mathrm{H}\{\cdot\}$ is defined in Eq. (3.64).

In this way, the analytic signal representation of FIBFs can be obtained. It should be noted that the obtained FIBFs from the signal are adaptive, complete, orthogonal, local, and uncorrelated. These FIBFs are based on the Fourier representation of the signal. The two frameworks namely, continuous FDM and discrete FDM are explained in following subsections.

## 8.6.2   CONTINUOUS FOURIER DECOMPOSITION METHOD

Consider a time-limited real-valued signal $x(t)$ defined in the interval $t \in [t_1,\, t_1 + T]$. For analysis purpose, periodic extension of $x(t)$ is considered, i.e., $x_T(t) = \sum_{k=-\infty}^{\infty} x(t - kT)$. The Fourier series expansion of $x_T(t)$ can be obtained by Eq. (2.21). The following expression is obtained by replacing the complex representation of $\cos(m\omega_0 t)$ and $\sin(m\omega_0 t)$ in Eq. (2.21):

$$x_T(t) = g_0 + \sum_{m=1}^{\infty} [g_m e^{jm\omega_0 t} + g_m^* e^{-jm\omega_0 t}] \tag{8.28}$$

where $g_m = (c_m - jd_m)$. Further, Eq. (8.28) can be written as,

$$x_T(t) = g_0 + \mathbb{R}\{z_T(t)\} \approx g_0 + \mathbb{R}\left\{\sum_{m=1}^{\infty} g_m e^{jm\omega_0 t}\right\} \tag{8.29}$$

The function $z_T(t)$ can be expressed in terms of $M$ AFM signals (analytic FIBFs) as,

$$z_T(t) = \sum_{i=1}^{M} a_i(t)e^{j\phi_i(t)} \tag{8.30}$$

In low-to-high frequency scan (LTH-FS), $a_1(t)e^{j\phi_1(t)} = \sum_{m=1}^{N_1} g_m e^{jm\omega_0 t}$, $a_2(t)e^{j\phi_2(t)} =$

$\sum_{m=N_1+1}^{N_2} g_m e^{jm\omega_0 t}, \dots, a_M(t)e^{j\phi_M(t)} = \sum_{m=N_{M-1}+1}^{\infty} g_m e^{jm\omega_0 t}$. Hence, from Eqs. (8.29) and

(8.30), we can express the analytic FIBFs in terms of complex Fourier coefficients as,

$$a_i(t)e^{j\phi_i(t)} = \sum_{m=N_{i-1}+1}^{N_i} g_m e^{jm\omega_0 t} \quad \text{for } i = 1, 2, \dots, M \qquad (8.31)$$

with $N_0 = 0$ and $N_M = \infty$. In order to obtain finite number of FIBFs, in LTH-FS, for each $i$, the limits of the summation in Eq. (8.31) starts with $N_{i-1} + 1$ and ends with $N_i$.

Similarly, for high-to-low frequency scan (HTL-FS), we obtain, $a_1(t)e^{j\phi_1(t)} =$

$\sum_{m=N_1}^{\infty} g_m e^{jm\omega_0 t}$, $a_2(t)e^{j\phi_2(t)} = \sum_{m=N_2}^{N_1-1} g_m e^{jm\omega_0 t}, \dots, a_M(t)e^{j\phi_M(t)} = \sum_{m=1}^{N_{M-1}-1} g_m e^{jm\omega_0 t}$ and

hence, we can express the analytic FIBFs in terms of complex Fourier coefficients as,

$$a_i(t)e^{j\phi_i(t)} = \sum_{m=N_i}^{N_{i-1}-1} g_m e^{jm\omega_0 t} \quad \text{for } i = 1, 2, \dots, M \qquad (8.32)$$

with $N_0 = \infty$ and $N_M = 1$. Based on the HTL-FS concept, the signal can be represented using finite number of analytic FIBFs.

The computation of $f_i(t) = \omega_{i(t)}/2\pi$ and $a_i(t)$ functions for each analytic FIBF of the signal $x(t)$ can help in constructing the Fourier Hilbert spectrum (FHS) based TFD $H_x(t, f)$. Further, the marginal FHS (MFHS) can be defined as,

$$H_x(f) = \int_0^T H_x(t, f) dt$$

Here, the MFHS denotes the measure of average AE for each frequency. In a similar way, the instantaneous energy density $E_x(t)$ can be defined as, $E_x(t) = \int_0^{f_M} H_x^2(t, f) df$, where, $f_M$ is the maximum frequency present in the signal.

## 8.6.3 DISCRETE FOURIER DECOMPOSITION METHOD

In real-world scenario, the sampled version of continuous-time signals are used for processing on computing devices; therefore, discrete FDM is needed to process such signals. The inverse DFT of a discrete-time signal $x(n)$ of length $N$ is defined as,

$$x(n) = \sum_{k=0}^{N-1} X(k)e^{\frac{j2\pi kn}{N}}$$

where, $X(k) = 1/N \sum_{n=0}^{N-1} x(n)e^{-j2\pi kn/N}$ is the DFT of the discrete-time signal $x(n)$.

Considering the length of the signal $N$ as an even number which leads to $X(0)$ and

$X(N/2)$ be real numbers. Under this consideration, the discrete-time signal $x(n)$ can be expressed as,

$$x(n) = X(0) + \underbrace{\sum_{k=1}^{N/2-1} X(k)e^{\frac{j2\pi nk}{N}}}_{z_1(n)} + X\left(\frac{N}{2}\right)e^{j\pi n} + \underbrace{\sum_{k=N/2+1}^{N-1} X(k)e^{\frac{j2\pi nk}{N}}}_{z_2(n)}$$

As $x(n)$ is a real-valued signal, $z_1(n)$ is complex conjugate of $z_2(n)$. Hence,

$$x(n) = X(0) + 2\text{Re}\{z_1(n)\} + X\left(\frac{N}{2}\right)(-1)^n \qquad (8.33)$$

where, $\text{Re}\{\cdot\}$ represents the real part of the complex value passed to it. Further, analytic representation $z_1(n)$ can be expressed as,

$$\sum_{k=1}^{N/2-1} X(k)e^{\frac{j2\pi kn}{N}} = \sum_{i=1}^{M} a_i(n)e^{j\phi_i(n)}$$

where, for the LTH-FS concept, the analytic FIBFs can be obtained as,

$$a_1(n)e^{j\phi_1(n)} = \sum_{k=1}^{N_1} X(k)e^{\frac{j2\pi kn}{N}}, \quad a_2(n)e^{j\phi_2(n)} = \sum_{k=N_1+1}^{N_2} X(k)e^{\frac{j2\pi kn}{N}},..., \quad a_M(n)e^{j\phi_M(n)} =$$

$\sum_{k=N_{M-1}}^{N/2-1} X(k)e^{\frac{j2\pi kn}{N}}$, where, $N_0 = 0$ and $N_M = N/2 - 1$. Minimum number of FIBFs can be obtained if we start frequency scan from $(N_{i-1}+1)$ to $(N/2-1)$ to obtain maximum value of $N_i$ such that the instantaneous phase $\phi_i(n)$ is monotonically increasing function, i.e.,

$$\{\omega_i(n) = \phi_i(n+1) - \phi_i(n)\} \geq 0, \forall n \qquad (8.34)$$

The estimated discrete IF represented by Eq. (8.34) is unbiased with zero group delay for linear frequency modulated signal $x(n)$.

Similarly, the FIBFs for discrete FDM based on the HTL-FS concept can be obtained as a continuous-time case.

The FDM method is a complete and data adaptive method which has the ability to localize frequency contents in time. In addition, the FDM method is suitable for analysis of nonlinear and non-stationary signals which together with Hilbert transform can provide TFR of the signals.

The decomposed components of an intrawave frequency modulated signal expressed by Eq. (8.35) obtained from FDM with LTH-FS and HTL-FS are shown in Fig. 8.10 and their corresponding TFDs are shown in Figs. 8.11 (a) and (b). For simulation, signal $x(t)$ expressed in Eq. (8.35) is sampled at sampling frequency $f_s = 8000$ Hz.

$$x(t) = \frac{2}{1.5 + \cos(2\pi t)} + \frac{\cos(30\pi t + 0.3\cos(64\pi t))}{1.8 + \sin(3\pi t)} \qquad (8.35)$$

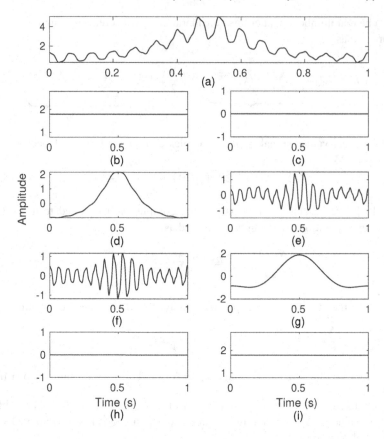

**Figure 8.10** (a) Signal $x(t)$, (b)–(e) the FIBFs obtained using FDM with LTH-FS concept, and (f)–(i) the FIBFs obtained using FDM with HTL-FS concept.

In order to determine fixed number of FIBFs, the FDM algorithm is implemented together with zero-phase filter bank concept. The FDM has been used for processing real-life signals [106, 107]. The FDM has been improved using the order-zero FBSE instead of Fourier decomposition and the resulting components are termed as Fourier-Bessel intrinsic band functions [108].

## 8.7  ITERATIVE EIGENVALUE DECOMPOSITION OF HANKEL MATRIX

The iterative eigenvalue decomposition of Hankel matrix (IEVDHM) is a data-adaptive signal decomposition technique which along with HSA provides TFR of the signal [109, 110, 111]. The IEVDHM-based decomposition of a non-stationary signal $x(n)$ into a set of AFM mono-component signals is defined with the help of following steps:

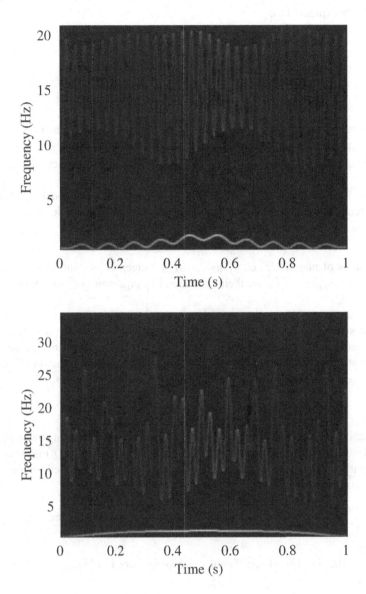

**Figure 8.11** (a) TFD of the signal $x(t)$ obtained from FDM: (a) LTH-FS concept and (b) HTL-FS concept.

Step 1: Obtaining Hankel matrix $H_K^x$ of size $K \times K$ from the signal $x(n)$ by considering the segments of length $K$ as column of it. The Hankel matrix can be mathematically expressed as,

$$
H_K^x = \begin{bmatrix} x(1) & x(2) & \cdots & x(K) \\ x(2) & x(3) & \cdots & x(K+1) \\ \vdots & \vdots & \ddots & \vdots \\ x(K) & x(K+1) & \cdots & x(N) \end{bmatrix}
$$

where, $K$ is known as embedding dimension of the Hankel matrix $H_K^x$ and $K = \frac{N+1}{2}$. For the signal $x(n)$, the embedding dimension $K$ can be chosen such that $K \geq \frac{f_s}{\Delta f}$, where $f_s$ denotes the sampling frequency and $\Delta f$ is the desired frequency resolution.

Step 2: The Hankel matrix $H_K^x$ is decomposed into a set of orthogonal eigenvectors and eigenvalues with the help of eigenvalue decomposition technique, which is expressed as,

$$
H_K^x = U \Lambda U^{\mathrm{T}}
$$

where, columns of matrix $U$, i.e., $u_1, u_2, \ldots, u_K$ are eigenvectors and diagonal elements of $\Lambda$, i.e., $\lambda_1, \lambda_2, \ldots, \lambda_K$ are their corresponding eigenvalues of the matrix $H_K^x$.

Step 3: The significant eigenvalue pairs $\{\lambda_i, \lambda_{K-i+1}\}$ of $H_K^x$ matrix are selected on the basis of significant threshold point (STP) criteria. The STP criteria states that the eigenvalues with magnitude greater than 10% of the maximum magnitude of eigenvalues are significant for analysis of the signal. The symmetric matrix $H_K^{x_i}$ corresponding to $i^{\mathrm{th}}$ significant eigenvalue pair $\{\lambda_i, \lambda_{K-i+1}\}$ is obtained as,

$$
H_K^{x_i} = u_i \lambda_i u_i^T + u_{K-i+1} \lambda_{K-i+1} u_{K-i+1}^T
$$

The signal component $x_i(n)$ associated with $H_K^{x_i}$ is obtained by computing mean of its $N$ skew diagonal elements. Then, the obtained $i^{\mathrm{th}}$ signal component $x_i(n)$ is checked whether it satisfies mono-component signal criteria (MSC) or not based on the following two conditions:

Condition 1: The difference between the number of zero-crossing and extrema is either zero or one.

Condition 2: The number of significant eigenvalue pairs in obtained signal component is one.

If $x_i(n)$ does not satisfy MSC, then, decomposition process is repeated for it.

Step 4: All the signal components satisfying MSC go under grouping step which states that, signal components are added together if their 1 dB bandwidth overlaps with each other. The resulting components after this step are considered as a set of AFM mono-component signals.

The decomposition of a 50 ms duration voiced speech signal (taken from CMU ARCTIC database having sampling frequency 16 kHz [31]) multiplied with a Hamming window of same duration using IEVDHM technique is shown in Fig. 8.12 and its corresponding TFD obtained using HSA of the decomposed components is shown in Fig. 8.13.

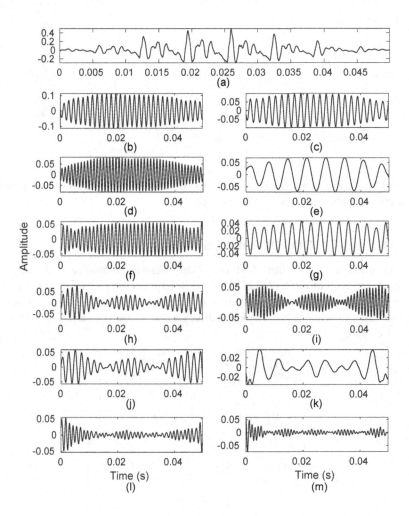

**Figure 8.12**   (a) Windowed voiced speech signal and (b)–(m) the decomposed signal components obtained using IEVDHM technique.

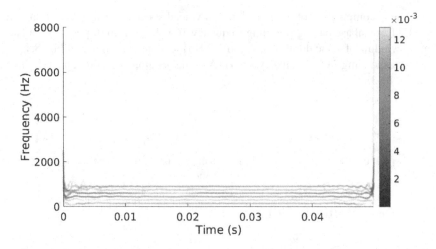

**Figure 8.13** TFD of windowed voiced speech signal based on HSA of the decomposed components from IEVDHM technique.

## 8.8 DYNAMIC MODE DECOMPOSITION

Dynamic mode decomposition (DMD) is a data-adaptive technique to decompose the signal into a set of modes [112, 113, 114]. It requires the collection of segments of length $K$ of the signal under analysis as it changes in time. These segments of the signal are used to generate two Hankel matrices $X$ and $X'$ each of the size $K \times (L-1)$ which are expressed as,

$$X = \begin{bmatrix} x(1) & x(2) & \cdots & x(L-1) \\ x(2) & x(3) & \cdots & x(L) \\ \vdots & \vdots & \ddots & \vdots \\ x(K) & x(K+1) & \cdots & x(K+L-2) \end{bmatrix}$$

$$X' = \begin{bmatrix} x(2) & x(3) & \cdots & x(L) \\ x(3) & x(4) & \cdots & x(L+1) \\ \vdots & \vdots & \ddots & \vdots \\ x(K+1) & x(K+2) & \cdots & x(K+L-1) \end{bmatrix}$$

The DMD algorithm is used for computation of eigenvalues and eigenvectors of the best-fit linear operator $T$ which maps the $i^{th}$ segment of the signal $x_i$ (where $i^{th}$ segment is defined as $x_i = [x(i), x(i+1), \ldots, x(i+K-1)]$) with the subsequent segment $x_{i+1}$ which can also be seen from the following expressions:

$$X' \approx TX \tag{8.36}$$

Mathematically, the operator $T$ is defined using the following expression:

$$T = \underset{T}{\operatorname{argmin}}||X' - TX||_F = X'X^{\dagger},    \tag{8.37}$$

where, $||.||_F$ represents the Frobenius norm operation and $^{\dagger}$ represents the pseudo-inverse operation.

For signal segment $x_i \in \mathbb{R}^K$, the number of elements in operator matrix $T$ is $K^2$, and computing its eigenvalues and eigenvectors may become difficult. The DMD helps in computing the significant eigenvalues and eigenvectors of $T$ without computing the operator matrix $T$. The steps involved in DMD algorithm are explained as follows:

Step 1: Compute the singular value decomposition (SVD) of matrix $X$ and then keep only $r$ significant singular values and their corresponding left and right singular vectors, which is defined as,

$$X \approx \tilde{U}\tilde{\Sigma}\tilde{V}^*    \tag{8.38}$$

where, $\tilde{U} \in \mathbb{C}^{K \times r}, \tilde{\Sigma} \in \mathbb{C}^{r \times r}, \tilde{V} \in \mathbb{C}^{(L-1) \times r}, r \leq (L-1)$, and $*$ denotes the conjugate transpose operation.

Step 2: The approximation matrix $T$ can be computed using pseudo-inverse of matrix $X$ as,

$$T = X'X^{\dagger} = X'\tilde{V}\tilde{\Sigma}^{-1}\tilde{U}^*    \tag{8.39}$$

However, we are only interested in $r$ significant eigenvalues and eigenvectors of $T$, hence, we may project $T$ onto $r$ significant left singular vectors of $X$ as,

$$\tilde{T} = \tilde{U}^*T\tilde{U} = \tilde{U}^*X'\tilde{V}\tilde{\Sigma}^{-1}    \tag{8.40}$$

Step 3: Spectral decomposition of matrix $\tilde{T}$ is defined as,

$$\tilde{T}W = W\Lambda    \tag{8.41}$$

Step 4: The DMD modes $\Phi$ are obtained using the following equation:

$$\Phi = X'\tilde{V}\tilde{\Sigma}^{-1}W    \tag{8.42}$$

These modes are eigenvectors of the approximation matrix $T$ corresponding to eigenvalues $\Lambda$.

DMD can be used to express the signals in terms of a data-adaptive spectral decomposition, which can be given as,

$$x_k = \Phi\Lambda^{k-1}b = \sum_{i=1}^{r} \phi_i\lambda_i^{k-1}b_i    \tag{8.43}$$

where, $\phi_i$ is the $i^{\text{th}}$ column vector of matrix $\Phi$, $\lambda_i$ is $i^{\text{th}}$ diagonal elements of diagonal matrix $\Lambda$, $b$ is mode amplitude and can be computed as $b = \Phi^{\dagger}x_1$, and $b_i$ is the $i^{\text{th}}$

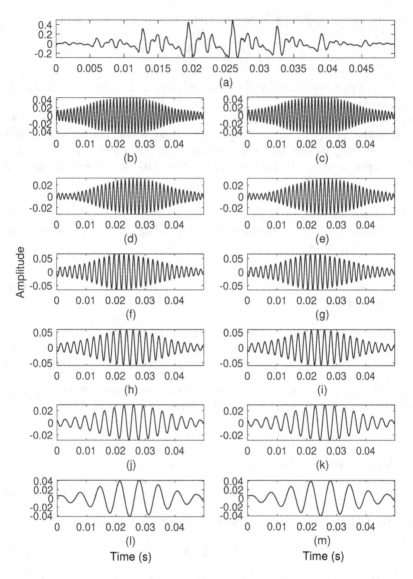

**Figure 8.14** (a) Windowed speech signal and (b)–(m) the decomposed signal components using DMD technique.

element of vector $b$. The matrix $\hat{X}$ can be obtained by representing $x_k$s as column vectors. The DMD modes can be obtained by performing diagonal averaging of $\hat{X}$ [72].

A 50 ms duration voiced speech signal is considered from CMU ARCTIC database with sampling frequency of 16 kHz [31]. The speech signal is multiplied with a Hamming window of same duration. The decomposition of the windowed speech signal based on DMD technique is shown in Fig. 8.14. The HSA can also be used on the decomposed components obtained from DMD technique in order to get TFR of the signal.

## PROBLEMS

Q 8.1 Compute the AM and FM bandwidths for various IMFs obtained using EMD method applied on any real-life signal. Comment on the nature of IMFs in terms of AM and FM modulations.

Q 8.2 Explain the effect on IF obtained from HTSA due to the addition of a constant to a pure sinusoidal signal.

Q 8.3 Consider any real-life signal and perform EMD, EWT, and VMD to obtain IMFs. For each case, plot Fourier spectrum of each IMF. Based on the observations of the Fourier spectrum of these IMFs, comment on the multiresolution property of all these three methods.

Q 8.4 Consider the signals expressed as follows:

(a) $x_1(n) = \cos(\omega_1 \frac{n}{f_s}) + \cos(\omega_2 \frac{n}{f_s}) + \cos(\omega_3 \frac{n}{f_s})$

(b) $x_2(n) = J_0\left(\frac{\omega_1 n}{f_s}\right) + J_0\left(\frac{\omega_2 n}{f_s}\right) + J_0\left(\frac{\omega_3 n}{f_s}\right)$

where, $f_s = 1000$ Hz, $\omega_1 = 100\pi$ rad/s, $\omega_2 = 200\pi$ rad/s, $\omega_3 = 300\pi$ rad/s. Apply FBSE-EWT and EWT on the above signals and compute AE and IF of all the components using HTSA. Compare the two methods based on the results.

Q 8.5 Consider the multicomponent signal,

$$x(n) = \cos\left[50\pi\frac{n}{f_s} + 10\cos(2\pi\frac{n}{f_s})\right] + \cos(100\pi\frac{n}{f_s})$$

which has a sampling frequency $f_s = 2$ kHz. Apply FBSE-EWT on the signal $x(n)$ using FBSE spectrum of order-zero and order-one. Also show the TFR for both the cases and comment on the obtained TFRs.

Q 8.6 Consider a multi-component signal,

$$x(n) = \cos[50\pi\frac{n}{f_s} + 10\cos(2\pi\frac{n}{f_s})] + [1 + \sin(4\pi\frac{n}{f_s})]\cos(200\pi\frac{n}{f_s})$$

Compute the TFRs of the signal $x(n)$ using HHT, EWT, and FBSE-EWT. Also obtain the TFR of each mono-component of the signal and add together in order to get the reference TFR of the signal. Compare the TFR of each of the three-techniques with reference TFR of the signal. Consider sampling frequency $f_s = 1000$ Hz for simulation.

Q 8.7 Take any EEG signal (from publicly available database) and decompose it using EMD method. Check whether mode-mixing problem is there or not by observing the IMFs and their corresponding spectrums. Add 10 dB white Gaussian noise to the same signal and compare the IMFs of original and noisy EEG signals. Apply EEMD on EEG signal and check if mode-mixing problem is solved or not.

Q 8.8 Consider a voiced speech signal of 25 ms duration. Compute its spectrogram and HSA based TFR of the IMFs obtained from VMD technique. Compare the results of spectrogram and VMD-HSA–based TFR in terms of the Renyi entropy measure. Refer [111] to get information about Renyi entropy measure.

Q 8.9 Consider any real-time ECG signal and perform FDM using LTH-FS and HTL-FS concepts in order to obtain the decomposed signal components. Plot and discuss the HSA–based TFR for both the cases.

Q 8.10 Take a pulse signal,

$$x(t) = \begin{cases} 1, & 1 \le t \le 2 \\ 0, & \text{otherwise} \end{cases}$$

Perform decomposition using EMD, VMD, EWT, FBSE-EWT, and FDM. Discuss the resolution of the decomposed components for each of the techniques, confirm if the multi-resolution property is captured or not. Sample the signal $x(t)$ using suitable sampling rate.

Q 8.11 For the signal,

$$x(t) = \cos\left[2\pi(100 + 15t)t\right] + 0.8\cos\left[2\pi(180 + 20t^2)t\right]$$

Plot the TFR of decomposed components before the grouping step and after the grouping step of IEVDHM method and discuss the need of grouping. Sample the signal $x(t)$ using suitable sampling rate.

Q 8.12 Show that the EMD is a nonlinear process. For any arbitrary AFM mono-component signal, compare AE and IF estimations using HTSA and NHT.

# 9 Applications

*"Every work has got to pass through hundreds of difficulties before succeeding. Those that persevere will see the light, sooner or later."* –Swami Vivekananda

## 9.1 OVERVIEW

Over the last few decades, TFA has experienced rapid growth and development in various applications. It is very difficult to find any area of science and technology where non-stationary signals are not present. Therefore, the application areas of TFA are very broad.

The application of TFA is well-known for data compression, signal denoising, and analysis of signals and systems in order to reduce storage memory requirement and enhancing the signal for further processing. Compression of biomedical signals finds application in the area of tele-medicine. TFA techniques find applications in analysis and classification of vibration signals in order to detect fault at early stage for machine diagnosis. Seismic signal analysis for earthquakes, landslides, and volcanic eruptions predication use TFA techniques.

TFA is a very useful tool for solving problems from various fields which include speech signal processing, biomedical signal processing, data compression, signal denoising, seismic signal processing, communication engineering, cognitive science, image processing, biomedical disease investigation, weather forecasting, chemical engineering, mechanical engineering, physics, pattern recognition, ocean engineering, etc. [115].

TFA method can be used to decompose multi-component non-stationary signals into a set of mono-component signals which can be considered simple and narrow-band in nature. The basis function design-based techniques such as STFT and wavelet transform require a prior design of basis functions in order to represent the signal in terms of multiple narrow-band signals. On the other hand, EMD was proposed for the analysis of non-stationary signals by adaptively decomposing them into various narrow-band components. Various adaptive signal decomposition techniques, motivated by EMD method, have been developed with proper mathematical formulation and provide improved decomposition. Analysis and interpretation of these simpler narrow-band signal components are easier and useful for feature extraction and TFR. Artificial intelligence-based classifiers can be developed using the extracted features from different components of the signals. 1D signal can be represented as an image based on TFR which can be used for classification of the signals based on deep learning classifiers. Representation of signal as an image has been found to be useful for biomedical signal classification.

Some representative applications of TFA have been explained in the following sections:

DOI: 10.1201/9781003367987-9

## 9.2 AUTOMATED DETECTION OF DISEASES USING BIOMEDICAL SIGNALS

Physiological signals like EEG, ECG, electromyogram (EMG), phonocardiogram (PCG), etc. provide meaningful information to monitor health conditions. Based on TFA techniques and artificial intelligence classifiers, automated diagnosis of various diseases such as, brain-related diseases, cardiac disorders, and muscular diseases can be carried out. The EEG signals have been used for diagnosis of neuro-disorder like epilepsy, Parkinson's disease, Alzheimer's disease, schizophrenia, sleep disorder, depression, etc [116, 117]. TFA method-based features help in classifying normal and abnormal classes of EEG signals corresponding to various disorders. The ECG signals which measure electrical activity of heart are used for detecting arrhythmia, cardiomyopathy, coronary artery disease, etc. TFA approaches have been utilized to develop automated identification systems for detection of heart-related diseases and neuromuscular diseases from ECG and EMG signals, respectively.

TFA methods help to extract significant features from the bio-medical signals. Mean frequency, standard deviation, signal energy in time-frequency plane, Hjorth parameters, entropies, etc. have been used as features for developing automated classification systems for the bio-medical signals corresponding to normal and abnormal categories [17, 108]. Two specific applications are described in the following sections:

### 9.2.1   EPILEPTIC SEIZURE DETECTION FROM EEG SIGNALS BASED ON EMD

Epilepsy is the most widely recognized and obliterating neurological disease experienced around the world. Epileptic patients experience seizure due to the abnormal neuronal firings, which may vary in symptoms [5, 118]. Few common symptoms are involuntary twitch of limbs, stare blankly, and loss of awareness. Depending on the region of the brain which gets affected, seizure can be categorized into two broad categories, namely focal and generalized seizure. Frequency of recurrent seizures may be controlled using proper treatment through medication, surgery, or vagus nerve stimulation, etc. But for that on-time diagnosis is necessary. The detection of epilepsy by visual examination of EEG signals is extremely tedious and might be erroneous, especially for long recordings. Based on the signal processing techniques, automated systems have been developed to monitor EEG signals for early diagnosis of epilepsy.

One of such methods for the detection of epileptic seizure is illustrated in the Fig. 9.1 (a). EEG signals are decomposed into narrow-band components (IMFs) using EMD. The AM bandwidth and FM bandwidth of the IMFs are used as features. The least-square support vector machine (LS-SVM) is used to classify EEG signals into seizure and non-seizure categories.

### 9.2.2   DETECTION OF CORONARY ARTERY DISEASE FROM ECG SIGNALS

Accumulation of fatty substances and cholesterol in the coronary arteries creates blockages and interrupts the normal flow of blood to the heart muscles. As a consequence of reduced blood flow, oxygen and other essential nutrients level will be

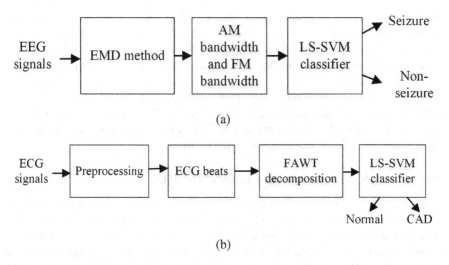

Figure 9.1    Block diagram of (a) epileptic seizure detection system and (b) CAD detection system.

lower than the normal level for functioning of heart muscles which weakens the heart muscles. Several complications are associated with coronary artery disease (CAD) like, heart attack, arrhythmia, heart failure. On-time treatment is helpful to avoid complications related to CAD. Generally, ECG signal is used for detection of CAD. However, contamination of artifact to ECG signals and subjective nature of visual investigation by expert reduce the classification accuracy. Advanced TFA technique-based methods have been developed to minimize the effect of artifacts and generate a bio-marker for the identification of CAD.

A FAWT-based method for CAD detection is described using the block diagram as shown in Fig. 9.1 (b). The ECG signals are segmented to extract the ECG beats containing P-wave, QRS complex, and T-waves. The FAWT is used to decompose the ECG beats. Cross-information potentials from FAWT-based detail coefficients have been computed as a set of features. The LS-SVM classifier is developed to classify ECG beats into normal and CAD categories with the help of extracted features [119].

## 9.3  DISEASE DETECTION AND DIAGNOSIS FROM BIOMEDICAL IMAGES

The 1D non-stationary signal analysis techniques such as DWT, EWT, FBSE-EWT, FBSE-FAWT, etc. have been extended for the decomposition of 2D bio-medical images wherein filtering operation is performed in row wise followed by column wise manner. For adaptive wavelet transform such as, EWT and FBSE-EWT methods where filter bank are designed from signal, the average of the spectrums of all rows and columns are needed to design the filter bank. X-ray, computed tomography (CT), magnetic resonance imaging (MRI), fundus images, ultrasound images, etc. are the most commonly used bio-medical images for diagnosis of diseases like, breast cancer, bone fractures and osteoporosis, Coronavirus disease (Covid-19) multiple sclerosis, brain injury from trauma, spinal cord disorders, tumors, gallbladder disease, breast lump, joint inflammation (synovitis), internal injuries, and internal bleeding [86].

Based on image decomposition methods, namely 2D-FBSE-EWT and 2D-FBSE-FAWT, automated techniques have been developed for the diagnosis of glaucoma and different grades of diabetic retinopathy (DR) and diabetic macular edema (DME), which have been discussed in the following sections:

### 9.3.1 GLAUCOMA DIAGNOSIS USING 2D-FBSE-EWT

Glaucoma is an eye disease in which fluid within the eye rises and puts pressure on optic nerves. This fluid pressure slowly damages the optic nerves, and if it is left untreated, it may lead to permanent vision loss. So the detection of glaucoma is necessary for on-time treatment. A computer-aided approach has been developed based on 2D-FBSE-EWT technique, which is shown in the block diagram of Fig. 9.2 (a). The 2D-FBSE-EWT decomposes the fundus images into sub-band images. Chip histogram features, gray level co-occurrence matrix (GLCM), and moment features are extracted from sub-band images to develop machine learning-based classifier to classify fundus images into normal and glaucoma categories [120].

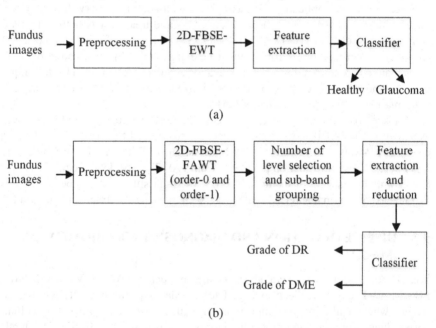

(a)

(b)

**Figure 9.2** Frameworks used for (a) automatic diagnosis of glaucoma using 2D-FBSE-EWT, (b) automatic diagnosis of different grades of DR and DME using 2D-FBSE-FAWT.

### 9.3.2  DIAGNOSIS OF DIABETIC RETINOPATHY AND DIABETIC MACULAR EDEMA USING 2D-FBSE-FAWT

In this section, a method for determining different grades of DR and DME is explained in Fig. 9.2 (b) [121]. In the preprocessing step, the contrast of images is improved using contrast limited adaptive histogram equalization (CLAHE). After pre-processing, order-zero and order-one 2D-FBSE-FAWT methods are used to decompose images into sub-band images. The level of decomposition is chosen automatically by setting a threshold criterion on the normalized energy of the approximation sub-band images. Several sub-band grouping techniques are used to group the sub-band images and local binary pattern and variance are extracted from these grouped sub-band images. Principle component analysis (PCA) is used to reduce the dimension of the features and based on machine learning classifier, different grades of DR and DME are determined.

## 9.4  EXTRACTION OF VITAL SIGNS FROM PHYSIOLOGICAL SIGNALS

The presence of noise and artifacts in the arterial blood pressure (ABP) waveform results in inaccurate analysis which in turn results in an inaccurate estimation of parameters and high false alarm rate. The delineation of ABP waveform and quality assessment can lead to better analysis. Further, TFA finds its application in extraction of heart rate (HR) and respiratory rate (RR) from photoplethysmography (PPG) signal [122].

### 9.4.1  BLOOD PRESSURE DELINEATION

For analysis, the ABP signal is first divided into small duration (4 seconds) window with 40% overlap. To reduce the edge effects, starting and ending 20% duration of the decomposed data are ignored for processing. Systolic blood pressure (SBP) estimation is performed by decomposing the signal into 7 IMFs using 10 iterations in sifting process of EEMD technique [123]. Then, SBP estimated signal is represented as $ABP_{IMF_{1-5}}$ which is obtained by the addition of first five IMFs. The SBP estimation performed here is of coarse type which involves the TFA technique in the framework. The coarse procedure involves detection of local maximas in $ABP_{IMF_{1-5}}$. Then, the low amplitude candidates are discarded using an adaptive threshold $TH_1$ which is 35% of average SBP amplitude in $ABP_{IMF_{1-5}}$. For onset detection, the sum of first 4 IMFs, i.e., $ABP_{IMF_{1-4}}$ are considered to be useful. The onset points detected using $ABP_{IMF_{1-4}}$ based on the estimated SBP involves the following steps:

Step 1: Find the minima in 60% of peak-to-peak interval duration prior to each SBP time estimate in $ABP_{IMF_{1-4}}$.

Step 2: Choose minima such that it is lower than 20% of average SBP amplitude in IMF domain and closest to SBP point.

In this way, one can estimate the SBP and onset points accurately from the PPG signal using the EEMD-based framework as discussed in aforementioned steps.

### 9.4.2   ESTIMATION OF HR AND RR FROM PPG SIGNAL USING EEMD AND PCA

In this application, firstly a single channel PPG signal is decomposed using EEMD technique [124, 125]. Then, the spectrum of the IMFs are obtained using FFT algorithm and then dominant frequency components were selected based on the presence of significant power in the IMFs. Then, IMFs having dominant frequencies outside the range [0.05, 2.5] Hz are discarded as they might be artifacts. Further, the IMFs having dominant frequencies in the range [0.05, 0.75] Hz have been selected for RR-group and having dominant frequencies in the range [0.75, 2.5] Hz are selected for HR-group.

In order to separate the HR and RR information, the PCA is applied on both the HR and RR groups. The first principal component (PC) obtained by applying PCA on HR and RR-groups represents the cardiac and respiratory activities, respectively. The frequency $f_{HR}$ is extracted from the first PC of HR group and similarly, $f_{RR}$ is extracted from the first PC of RR group. Finally, the HR and RR are computed using the following expressions:

$$HR = 60 \times f_{HR} \text{ (beats/min)} \tag{9.1}$$

$$RR = 60 \times f_{RR} \text{ (beats/min)} \tag{9.2}$$

In this manner, the HR and RR intervals are estimated successfully from PPG signal and the block diagram of the same is shown in Fig. 9.3.

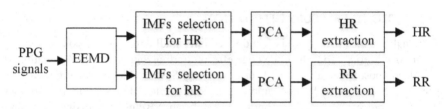

**Figure 9.3**   HR and RR estimations from PPG signals.

### 9.5   BRAIN-COMPUTER INTERFACE

Brain-computer interface (BCI) is a system that uses signals generated from brain for direct control of physical objects without using muscular activity or body movements [126, 127]. The brain activity responses are captured through EEG. The EEG signals acquired from human brain are non-stationary in nature thus, to extract information about brain states during different mental tasks, TFA techniques can be helpful. The adaptive data decomposition techniques such as EMD and VMD are used to enhance the quality of the EEG signals. The STFT and wavelet transform-based TFRs have been used to classify EEG signals using convolutional neural networks [128]. Several features can be extracted from biomedical signals based on the TFA techniques for

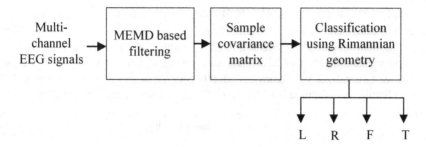

**Figure 9.4** MEMD-based MI-BCI classification technique.

developing artificial intelligence-based classifier to derive controlling command for external devices. The BCI systems have wide range of applications which include rehabilitation, educational and self-regulation, security and authentication, games and entertainment, etc.

Motor imagery (MI) BCI (MI-BCI) is one such application where imaginary motor movement is predicted from EEG signal. A multivariate EMD (MEMD) based method for MI-BCI is described using a block diagram shown in Fig. 9.4. In MEMD-based filtering step, a multi-channel EEG signal is decomposed into multivariate IMFs (MIMFs). Based on the mean frequency, specific MIMFs are chosen to perform filtering which enhances the quality of EEG. Sample covariance matrix of the enhanced EEG signal is used as feature to develop classifier. The Riemannian geometry framework is used for classification of EEG signals into different MI tasks namely, left hand (L), right hand (R), foot (F), and tongue (T) imagery movements.

## 9.6 TFA FOR SPEECH PROCESSING

As speech is non-stationary signal, the TFA-based techniques can improve the performance of speech processing systems which are developed based on time-domain or frequency-domain techniques. TFA techniques have been used for speech analysis, synthesis, enhancement, coding, compression, feature extraction, and classification for various applications. The following two subsections explain the applications of TFA technique on speech recognition and pitch determination:

### 9.6.1 ROBUST AM-FM FEATURES FOR SPEECH RECOGNITION

The speech signal depicts a non-stationary behavior. Hence, the AFM model-based features are found to be helpful in speech recognition system [129]. These features attempt to capture information which are not observed in linear source-filter model. Later, these features after embedding with mel-frequency cepstral coefficients (MFCCs) are used for training and testing of speech recognition system.

In AFM feature extraction, the AE and IF functions of each resonance signal $s_k(t)$ are extracted. The resonance signals are extracted using a six-filter in mel-spaced Gabor filter bank. The Gabor filters are used due to their optimal time-frequency discriminability nature. In the filter bank, 50% overlapping between the consecutive filters are considered. After designing the filters, the resonance signals $s_k(t)$ are extracted. Then, a binomial smoothing of energy signal is performed to overcome the high-pass modeling error of DESA-1 algorithm. The IF $f_k(t)$ and AE $a_k(t)$ are then computed using DESA-1 algorithm. Finally, the frequency modulation percentage (FMP) feature is computed for each $s_k(t)$ which is mathematically expressed as,

$$\text{FMP}_k = \frac{B_k}{F_k}, \text{ where } 1 \leq k \leq 6 \tag{9.3}$$

where,

$$F_k = \frac{\int_0^T f_k(t) a_k^2(t) dt}{\int_0^T a_k^2(t) dt} \tag{9.4}$$

and

$$B_k = \frac{\int_0^T \left[ \dot{a}_k^2(t) + (f_k(t) - F_k)^2 a_k^2(t) \right] dt}{\int_0^T a_k^2(t) dt} \tag{9.5}$$

$T$ represents the length of the window considered for short duration feature extraction and $\dot{a}_k(t) = \frac{da_k(t)}{dt}$. The cepstral mean subtraction is applied on MFCC features. The post filtering using median filter is also employed to AE and IF functions in order to have robust estimation. After using these features, the performance of speech recognition system has been found to be improved as compared to the MFCC features.

### 9.6.2   PITCH DETERMINATION USING HHT

In order to obtain TFD, the HHT consists of two main steps: the first one is EMD which decomposes the signal into IMFs and second one is HTSA which provides IF and AE functions [130]. The joint representation of IF and AE in time-frequency plane provides the TFD. It should be noted that the analysis of speech signals containing entire frequency range is not required for pitch determination application, as the information about formants are of no use in this case. Hence, the speech signal is first low-pass filtered using a 800 Hz sharp cutoff filter and then the resultant signal is decomposed into IMFs using EMD technique. After this, computation of the IF and AE of each of the IMF using HTSA is carried out. The values of IFs which are outside the range 60–500 Hz or variation is higher than 100 Hz within 5 ms duration have been replaced by zero. In addition, the IFs whose amplitudes are less than 10% of maximum amplitude are set to zero. After these processings, the remaining fundamental frequency components are merged together to obtain single pitch frequency of the speech signal. The HHT-based method for pitch determination improves the accuracy of pitch estimation and resolution of the pitch.

### 9.7   APPLICATIONS IN COMMUNICATION ENGINEERING

In communication engineering, the modulated signals contain time-varying information which can be represented in time-frequency domain in order to have

**Figure 9.5**   Time-frequency domain-based mode identification.

representation for better extraction of information. Wide range of applications of TFA in communication engineering includes radar signal analysis, mitigation of jammer, interference excision, and channel identification. Two such applications of TFA are described in the following sections:

### 9.7.1   MODE IDENTIFICATION IN WIRELESS SOFTWARE-DEFINED RADIO

The time-frequency domain-based model for identification of mode in wireless software-defined radio is shown in Fig. 9.5 [131]. First the received analog signal is converted into digital signal using analog to digital converter (ADC). Then, the TFA is performed on the signal to obtain TFR. For this purpose, Choi-William distribution has been studied.

From the obtained TFR, the two features namely, spectral spread (SS) and maximum duration (MD) of the signal are computed. The TFR is made binary by using threshold. Threshold is selected empirically based on trial and error basis. Classification has been done based on the aforementioned two features to classify data into four classes, namely, noise, bluetooth and additive white Gaussian noise (AWGN), wireless local area network (WLAN) and AWGN, and WLAN and bluetooth in AWGN.

### 9.7.2   JAMMER MITIGATION

Another application of TFA in communication system has been observed in global navigation satellite system (GNSS) receivers [132]. The jamming signals from various privacy devices interfere with GNSS signals thereby degrading the performance of the receiver. To remove these jamming signals, they are first estimated using IF and chirp rate estimates obtained from TFA. These estimated parameters are fed into a Kalman filter which estimates the jamming signal. This estimated signal is then passed through a subtractor to complete the jammer supression and obtain clean received signal. The general block diagram for jammer mitigation system based on Kalman filter is shown in Fig. 9.6.

## 9.8   POWER QUALITY ASSESSMENT

TFA techniques are being used for disturbance identification, localization, and classification in power systems. For healthy and economic operation of power system, it is imperative to detect, diagnose, and mitigate different types of disturbances like sag, swell, harmonics, transients etc. As most of the signals are non-stationary in nature, hence TFA techniques are useful for analyzing these disturbances. Various

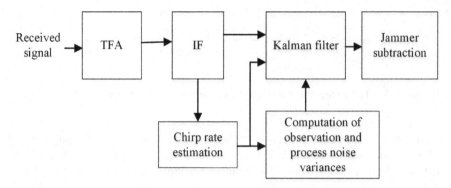

**Figure 9.6**  A general block diagram for jammer mitigation system.

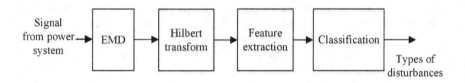

**Figure 9.7**  Block diagram of PQ assessment using EMD.

time-frequency domain methods have been used. One such method for assessment of power quality based on EMD method is described in Fig. 9.7 [133].

The disturbance signals are decomposed into number of IMFs using EMD, then Hilbert transform is utilized for feature extraction from each IMF. Energy of the analytic IMFs, standard deviation (SD) of AE, and SD of instantaneous phase of the IMFs are extracted as features to train probabilistic neural network machine learning classifiers to assess the power quality.

## 9.9  MACHINERY FAULT DIAGNOSIS

Machines are an integral part of any industrial operation. They contain various rotatory and translatory moving parts such as flywheel, spindles, shafts, etc. Fault in any of these parts can hamper the whole process which may result in high downtime and capital loss. One such part which is used to control the speed of the machines is known as gear box. It is one of the important parts of the rotating machines. Pitting, chipping, and crack are the typical faults that occurs in gear box. These faults are analyzed using the vibration signal of the machines because vibration signal may contain amplitude and phase modulations generated due to faults. These vibration signals are non-stationary in nature and hence can be analyzed by exploiting various TFA methods.

One such method uses WPT along with the Hilbert transform for gear fault diagnosis [134]. The acquired vibration signal is first pre-processed to remove the effects

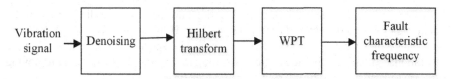

**Figure 9.8**   WPT-based gear fault diagnosis technique.

of noise. Hilbert transform is then applied to the noise free signal which extracts AE of the acquired signal. After discarding the carrier frequencies, the obtained envelope contains the modulation information which can be further analyzed for fault detection. Now this non-stationary envelope is analyzed using WPT to detect modulating signals. The reconstructed signals are obtained using WPT coefficients which are used to extract characteristic frequency representing the type of local fault in the gear box. Figure 9.8 presents the block diagram of the above described method.

## 9.10   CHEMICAL ENGINEERING

For enhancing the safety, quality control, productivity, improved methods for fault diagnosis, identification, modeling of chemical process systems are essential. One of the challenging steps is extracting meaningful information from the time-varying sensory information. TFA is an emerging area which can provide useful interpretation of the data. Process controls, numerical solutions to differential equations, process signal feature analysis are a few applications of TFA techniques in the domain of chemical engineering. Two such applications of TFA are described in the following sections.

### 9.10.1   PREDICTION OF DIFFERENT PATTERNS IN IVR

The main issues in chemical kinetics are intra-molecular vibrational randomization (IVR) processes, their rate, and the degree of completeness. To study the time-varying nature of the IVR pattern for different initial energies, the STFT of the time correlation of physical data like dipole moment, has been utilised. The change in spectral energy is observed from the STFT to predict the different patterns in IVR [135].

### 9.10.2   APPLICATION IN UV SPECTROSCOPY

In the laboratory, the various instruments or devices are used for measuring different signals like, chromatogram, electropherogram, and voltammogram. Analysis of these signals are useful for studying the properties of chemicals and pharmaceuticals. UV spectrophotometric method is a low-cost and acceptable alternative of chromatogram but it faces the problem of spectral overlapping [136].

The multi-component mixture (both active and inactive ingredients of sample) analysis uses the concept of TFA techniques like, Hilbert transform, STFT, CWT to overcome the effect of spectral interference in UV spectrophotometry. Also, these techniques are applied in the pre-treatment step to simplify the signals and removing the measurement error during signal spectral analysis.

## 9.11　FINANCIAL APPLICATIONS

The analysis of financial time series using frequency-domain analysis is relatively less explored to predict the stock market, oil market, and business cycle [137, 138, 139]. The stock market is influenced by a variety of factors at different time periods, such as, it has dynamic frequency characteristics which can be explored to analyze the stock market [140]. As this data is non-stationary in nature, the Fourier transform fails to provide cyclical characteristics of the data.

In stock market prediction, the TFA is used to extract features which are fed to the classifier. This data is analyzed using the Gabor transformation, STFT, WVD, etc. The fear of investors and expectations in stock markets of BRICS economies have also been studied using CWT, wavelet transform coherence phase difference, and wavelet multiple correlation [141].

The relationship between oil price, stock returns, and exchange rate have been studied using TFA wherein, wavelet analysis is used. In such cases, methods such as cross wavelet spectrum and wavelet coherence aid in finding the correlation between two different non-stationary time series data. The cyclical behavior of business cycle is also studied from the time-domain and frequency-domain perspectives. This analysis is conducted using spectrum estimate via periodogram and autoregressive process.

For example, a method for stock price prediction using TFA and convolution neural network has been developed, where the original time series data are transformed into a 2D time-frequency feature using STFT. A filter bank consisting of triangular filters is used for feature extraction. These features represent the time localized frequency information which are fed to a convolution neural network to predict the stock price.

## 9.12　OCEAN ENGINEERING

Ocean engineering finds applications in developing monitoring and controlling system in coastal environment. In this field of engineering, TFA techniques are found to be applicable for analyzing non-stationary signals to find the effect of ship wake on different structures situating near coast area, predicting tide level, assessment of fatigue damage of offshore structure, etc [142, 143, 144].

TFA is performed on the cross-section of ship wake and visualization can be done in order to facilitate the surveillance of unmonitored vessels, remote sensing, and quantifying the adverse effects of propagating waves in coastal zone. Spectrogram is used to classify the resulting data into sub-critical or super-critical flow with different wave patterns formed for each case.

Prediction of tide-level is important for construction planning of coastal harbors, structures, and biological resource management. Tidal wave signal can be expressed as superposition of time-varying sinusoids. The CWT is used to represent the tidal wave in time-frequency domain in order to extract harmonic (or quasi harmonic) components. From frequency, AE, phase of tidal signals can be used to predict the tidal level of the next hour.

# References

1. John A Stuller. *An Introduction to Signals and Systems.* Thomson, 2007.

2. Michael J Roberts. *Fundamentals of Signals and Systems.* McGraw-Hill Science/Engineering/Math, 2007.

3. Ary L Goldberger, Luis AN Amaral, Leon Glass, Jeffrey M Hausdorff, Plamen Ch Ivanov, Roger G Mark, Joseph E Mietus, George B Moody, Chung-Kang Peng, and H Eugene Stanley. Physiobank, physiotoolkit, and physionet: components of a new research resource for complex physiologic signals. *Circulation*, 101(23):e215–e220, 2000.

4. George B Moody and Roger G Mark. The impact of the MIT-BIH arrhythmia database. *IEEE Engineering in Medicine and Biology Magazine*, 20(3):45–50, 2001.

5. Ralph G Andrzejak, Klaus Lehnertz, Florian Mormann, Christoph Rieke, Peter David, and Christian E Elger. Indications of nonlinear deterministic and finite-dimensional structures in time series of brain electrical activity: Dependence on recording region and brain state. *Physical Review E*, 64(6):061907, 2001.

6. Bhagwandas Pannalal Lathi and Roger A Green. *Linear Systems and Signals*, volume 2. Oxford University Press, New York, 2005.

7. Gerwin Schalk, Dennis J McFarland, Thilo Hinterberger, Niels Birbaumer, and Jonathan R Wolpaw. BCI2000: a general-purpose brain-computer interface (BCI) system. *IEEE Transactions on Biomedical Engineering*, 51(6):1034–1043, 2004.

8. Alan V Oppenheim, Alan S Willsky, and Syed Hamid Nawab. *Signals & Systems.* Pearson Indian Education, 2015.

9. Dwight F Mix and Kraig J Olejniczak. *Elements of Wavelets for Engineers and Scientists.* John Wiley & Sons, 2003.

10. Matthew NO Sadiku and Warsame Hassan Ali. *Signals and Systems: A Primer with MATLAB.* CRC Press, 2015.

11. Gabriele D'Antona and Alessandro Ferrero. *Digital Signal Processing for Measurement Systems: Theory and Applications.* Springer Science & Business Media, 2005.

12. Ali N Akansu and Richard A Haddad. *Multiresolution Signal Decomposition: Transforms, Subbands, and Wavelets.* Academic Press, 2001.

13. Gordon E Carlson. *Signal and Linear System Analysis.* John Wiley Hoboken, NJ, 1998.

14. Peter Bloomfield. *Fourier Analysis of Time Series: An Introduction.* John Wiley & Sons, 2004.

15. Eric W Hansen. *Fourier Transforms: Principles and Applications*. John Wiley & Sons, 2014.

16. Rajendra Prasad Singh and Sadanand Damodar Sapre. *Communication Systems, 2E.* Tata McGraw-Hill Education, 2008.

17. Svend Gade and Henrik Herlufsen. Use of weighting functions in DFT/FFT analysis (part i). *Brüel & Kjær Technical Review*, 3:1–28, 1987.

18. Yeong Ho Ha and John A Pearce. A new window and comparison to standard windows. *IEEE Transactions on Acoustics, Speech, and Signal Processing*, 37(2):298–301, 1989.

19. Fredric J Harris. On the use of windows for harmonic analysis with the discrete Fourier transform. *Proceedings of the IEEE*, 66(1):51–83, 1978.

20. Albert Nuttall. Some windows with very good sidelobe behavior. *IEEE Transactions on Acoustics, Speech, and Signal Processing*, 29(1):84–91, 1981.

21. Pradeep Kumar Chaudhary, Vipin Gupta, and Ram Bilas Pachori. Fourier-Bessel representation for signal processing: A review. *Digital Signal Processing*, DOI: https://doi.org/10.1016/j.dsp.2023.103938, 2023.

22. George B Arfken and Hans J Weber. *Mathematical Methods for Physicists*. American Association of Physics Teachers, 1999.

23. Norman William McLachlan. *Bessel functions for Engineers*. University of Illinois, Oxford University Press, London, England, 1961.

24. Kaliappan Gopalan, Timothy R Anderson, and Edward J Cupples. A comparison of speaker identification results using features based on cepstrum and Fourier-Bessel expansion. *IEEE Transactions on Speech and Audio Processing*, 7(3):289–294, 1999.

25. Ram Bilas Pachori and Pradip Sircar. EEG signal analysis using FB expansion and second-order linear TVAR process. *Signal Processing*, 88(2):415–420, 2008.

26. Ram Bilas Pachori and Pradip Sircar. Non-stationary signal analysis: Methods based on Fourier-Bessel representation. *LAP LAMBERT Academic Publishing, Germany*, 630, 2010.

27. Jim Schroeder. Signal processing via Fourier-Bessel series expansion. *Digital Signal Processing*, 3(2):112–124, 1993.

28. Erwin Kreyszig. *Advanced Engineering Mathematics, 10th Edition*. Wiley, 2009.

29. Alexander D Poularikas. *Handbook of Formulas and Tables for Signal Processing*. CRC press, 2018.

30. Ram Bilas Pachori and Pradip Sircar. Analysis of multi-component non-stationary signals using Fourier-Bessel transform and Wigner distribution. In *2006 14th European Signal Processing Conference*, pages 1–5. IEEE, 2006.

31. John Kominek and Alan W Black. The CMU Arctic speech databases. In *Fifth ISCA Workshop on Speech Synthesis*, 2004.

32. Petros Maragos, James F Kaiser, and Thomas F Quatieri. Energy separation in signal modulations with application to speech analysis. *IEEE Transactions on Signal Processing*, 41(10):3024–3051, 1993.

33. Ram Bilas Pachori and Pradip Sircar. Analysis of multicomponent AM-FM signals using FB-DESA method. *Digital Signal Processing*, 20(1):42–62, 2010.

34. Pradip Sircar and Rakesh K Saini. Parametric modeling of speech by complex AM and FM signals. *Digital Signal Processing*, 17(6):1055–1064, 2007.

35. Pradip Sircar and Mohanjeet Singh Syali. Complex AM signal model for non-stationary signals. *Signal Processing*, 53(1):35–45, 1996.

36. Pradip Sircar and Sanjay Sharma. Complex FM signal model for non-stationary signals. *Signal Processing*, 57(3):283–304, 1997.

37. Mohan Bansal and Pradip Sircar. A novel AFM signal model for parametric representation of speech phonemes. *Circuits, Systems, and Signal Processing*, 38:4079–4095, 2019.

38. Avinash Shrikant Hood, Ram Bilas Pachori, Varuna Kumar Reddy, and Pradip Sircar. Parametric representation of speech employing multi-component AFM signal model. *International Journal of Speech Technology*, 18(3):287–303, 2015.

39. Leon Cohen. *Time-Frequency Analysis*. Prentice Hall, New Jersey, 1995.

40. Martin Vetterli, Jelena Kovačević, and Vivek K Goyal. *Foundations of Signal Processing*. Cambridge University Press, 2014.

41. Vaclav Cizek. Discrete Hilbert transform. *IEEE Transactions on Audio and Electroacoustics*, 18(4):340–343, 1970.

42. Patrick J Loughlin and Berkant Tacer. Comments on the interpretation of instantaneous frequency. *IEEE Signal Processing Letters*, 4(5):123–125, 1997.

43. Alexandros Potamianos and Petros Maragos. A comparison of the energy operator and the Hilbert transform approach to signal and speech demodulation. *Signal Processing*, 37(1):95–120, 1994.

44. Patrick Flandrin. *Time-Frequency/Time-Scale Analysis*. Academic Press, 1998.

45. Patrick Flandrin. *Explorations in time-Frequency Analysis*. Cambridge University Press, 2018.

46. Thomas F Quatieri. *Discrete-Time Speech Signal Processing: Principles and Practice*. Pearson Education India, 2006.

47. Dennis Gabor. Theory of communication. part 1: The analysis of information. *Journal of the Institution of Electrical Engineers-Part III: Radio and Communication Engineering*, 93(26):429–441, 1946.

48. Stéphane Mallat. *A Wavelet Tour of Signal Processing*. Elsevier, 2009.

49. Jordi Burriel-Valencia, Ruben Puche-Panadero, Javier Martinez-Roman, Angel Sapena-Bano, and Manuel Pineda-Sanchez. Short-frequency Fourier transform for fault diagnosis of induction machines working in transient regime. *IEEE Transactions on Instrumentation and Measurement*, 66(3):432–440, 2017.

50. Jont B Allen and Lawrence R Rabiner. A unified approach to short-time Fourier analysis and synthesis. *Proceedings of the IEEE*, 65(11):1558–1564, 1977.

51. Alfred Mertins. *Signal Analysis: Wavelets, Filter Banks, Time-Frequency Transforms and Applications*. John Wiley & Sons, Inc., 1999.

52. Alan V Oppenheim, Ronald W Schafer, and John R Buck. *Discrete-Time Signal Processing. Vol. 2*. Upper Saddle River, NJ: Prentice Hall, 2001.

53. Lawrence R Rabiner and Ronald W Schafer. *Digital Processing of Speech Signals*. Pearson Education India, 1978.

54. Ingrid Daubechies. *Ten Lectures on Wavelets*. SIAM, 1992.

55. Raghuveer M Rao and Ajit S Bopardikar. *Wavelet Transforms: Introduction to Theory and Applications*. Pearson Education India, 1998.

56. Agostino Abbate, Casimer M DeCusatis, and Pankaj K Das. *Wavelets and Subbands: Fundamentals and Applications*. Springer Science & Business Media, 2012.

57. Ingrid Daubechies. The wavelet transform, time-frequency localization and signal analysis. *IEEE Transactions on Information Theory*, 36(5):961–1005, 1990.

58. Gilbert Strang and Truong Nguyen. *Wavelets and Filter Banks*. Wellesley-Cambridge Press, 1996.

59. Martin Vetterli and Cormac Herley. Wavelets and filter banks: Theory and design. *IEEE Transactions on Signal Processing*, 40:2207–2232, 1992.

60. Jelena Kovacevic and Martin Vetterli. *Wavelets and Subband Coding*. Prentice Hall PTR, 1995.

61. Ronald L Allen and Duncan W Mills. *Signal Analysis: Time, Frequency, Scale, and Structure*. John Wiley & Sons, 2004.

62. NG De Bruijn. A theory of generalized functions, with applications to Wigner distribution and Weyl correspondence. *Nieuw Archief voor Wiskunde*, 21(3):205–280, 1973.

63. Jean Ville. Theorie et application dela notion de signal analytique. *Câbles et Transmissions*, 2(1):61–74, 1948.

64. Boualem Boashash. *Time-Frequency Signal Analysis and Processing: A Comprehensive Reference*. Academic Press, 2015.

65. TACM Claasen and Mecklenbräuker WFG. The Wigner distribution – a tool for time-frequency signal analysis, part I: Continuous-time signals. *Philips Journal of Research*, 35:217–250, 1980.

66. TACM Claasen and Mecklenbräuker WFG. The Wigner distribution – a tool for time-frequency signal analysis, part II: Discrete-time signals. *Philips Journal of Research*, 35:276–300, 1980.

67. TACM Claasen and Mecklenbräuker WFG. The Wigner distribution – a tool for time-frequency signal analysis, part III: Relations with other time-frequency signal transformations. *Philips Journal of Research*, 35:372–389, 1980.

68. Ram Bilas Pachori and Anurag Nishad. Cross-terms reduction in the Wigner–Ville distribution using tunable-Q wavelet transform. *Signal Processing*, 120:288–304, 2016.

69. Rishi Raj Sharma, Avinash Kalyani, and Ram Bilas Pachori. An empirical wavelet transform-based approach for cross-terms-free Wigner–Ville distribution. *Signal, Image and Video Processing*, 14(2):249–256, 2020.

70. Yunzi Chai and Xudong Zhang. EMD-WVD time-frequency distribution for analysis of multi-component signals. In *Fourth International Conference on Wireless and Optical Communications*, volume 9902, pages 190–196. SPIE, 2016.

71. Rishi Raj Sharma and Ram Bilas Pachori. Improved eigenvalue decomposition-based approach for reducing cross-terms in Wigner–Ville distribution. *Circuits, Systems, and Signal Processing*, 37(8):3330–3350, 2018.

72. Vivek Kumar Singh and Ram Bilas Pachori. Sliding eigenvalue decomposition-based cross-term suppression in Wigner–Ville distribution. *Journal of Computational Electronics*, 20(6):2245–2254, 2021.

73. Rishi Raj Sharma, Preeti Meena, and Ram Bilas Pachori. Enhanced time–frequency representation based on variational mode decomposition and Wigner–Ville distribution. In *Recent Trends in Image and Signal Processing in Computer Vision*, pages 265–284. Springer, 2020.

74. Ram Bilas Pachori and Pradip Sircar. A new technique to reduce cross terms in the Wigner distribution. *Digital Signal Processing*, 17(2):466–474, 2007.

75. Shie Qian and Dapang Chen. Decomposition of the Wigner-Ville distribution and time-frequency distribution series. *IEEE Transactions on Signal Processing*, 42(10):2836–2842, 1994.

76. SV Narasimhan and Malini B Nayak. Improved Wigner–Ville distribution performance by signal decomposition and modified group delay. *Signal Processing*, 83(12):2523–2538, 2003.

77. Farook Sattar and Goran Salomonsson. The use of a filter bank and the Wigner-Ville distribution for time-frequency representation. *IEEE Transactions on Signal Processing*, 47(6):1776–1783, 1999.

78. Rohan Panda, Sahil Jain, RK Tripathy, Rishi Raj Sharma, and Ram Bilas Pachori. Sliding mode singular spectrum analysis for the elimination of cross-terms in Wigner–Ville distribution. *Circuits, Systems, and Signal Processing*, 40(3):1207–1232, 2021.

79. Paulo Gonçalves and Richard G Baraniuk. Pseudo affine Wigner distributions: Definition and kernel formulation. *IEEE Transactions on Signal Processing*, 46(6):1505–1516, 1998.

80. Gregorio Andria and Mario Savino. Interpolated smoothed pseudo Wigner-Ville distribution for accurate spectrum analysis. *IEEE Transactions on Instrumentation and Measurement*, 45(4):818–823, 1996.

81. H-I Choi and William J Williams. Improved time-frequency representation of multicomponent signals using exponential kernels. *IEEE Transactions on Acoustics, Speech, and Signal Processing*, 37(6):862–871, 1989.

82. Patrick Flandrin and Olivier Rioul. Affine smoothing of the Wigner-Ville distribution. In *International Conference on Acoustics, Speech, and Signal Processing*, pages 2455–2458. IEEE, 1990.

83. José E Moyal. Quantum mechanics as a statistical theory. In *Mathematical Proceedings of the Cambridge Philosophical Society*, volume 45, pages 99–124. Cambridge University Press, 1949.

84. Ronald R Coifman, Yves Meyer, Steven Quake, and M Victor Wickerhauser. Signal processing and compression with wavelet packets. In *Wavelets and Their Applications*, pages 363–379. Springer, 1994.

85. Gary G Yen and K-C Lin. Wavelet packet feature extraction for vibration monitoring. *IEEE Transactions on Industrial Electronics*, 47(3):650–667, 2000.

86. Pradeep Kumar Chaudhary and Ram Bilas Pachori. FBSED based automatic diagnosis of COVID-19 using X-ray and CT images. *Computers in Biology and Medicine*, 134:104454, 2021.

87. Ingrid Daubechies, Jianfeng Lu, and Hau-Tieng Wu. Synchrosqueezed wavelet transforms: An empirical mode decomposition-like tool. *Applied and Computational Harmonic Analysis*, 30(2):243–261, 2011.

88. Ilker Bayram and Ivan W Selesnick. Frequency-domain design of overcomplete rational-dilation wavelet transforms. *IEEE Transactions on Signal Processing*, 57(8):2957–2972, 2009.

89. Ivan W Selesnick. Wavelet transform with tunable Q-factor. *IEEE Transactions on Signal Processing*, 59(8):3560–3575, 2011.

90. Ivan W Selesnick. TQWT toolbox guide. *Electrical and Computer Engineering, Polytechnic Institute of New York University. Available online at: http://eeweb.poly. edu/iselesni/TQWT/index. html*, 2011.

91. Shivnarayan Patidar, Ram Bilas Pachori, and Niranjan Garg. Automatic diagnosis of septal defects based on tunable-Q wavelet transform of cardiac sound signals. *Expert Systems with Applications*, 42(7):3315–3326, 2015.

92. Ilker Bayram. An analytic wavelet transform with a flexible time-frequency covering. *IEEE Transactions on Signal Processing*, 61(5):1131–1142, 2012.

93. Vipin Gupta and Ram Bilas Pachori. Classification of focal EEG signals using FBSE based flexible time-frequency coverage wavelet transform. *Biomedical Signal Processing and Control*, 62:102124, 2020.

94. Ivan W Selesnick, Richard G Baraniuk, and Nick C Kingsbury. The dual-tree complex wavelet transform. *IEEE Signal Processing Magazine*, 22(6):123–151, 2005.

95. Ivan W Selesnick. Hilbert transform pairs of wavelet bases. *IEEE Signal Processing Letters*, 8(6):170–173, 2001.

96. Nick Kingsbury. The dual-tree complex wavelet transform: a new efficient tool for image restoration and enhancement. In *9th European Signal Processing Conference (EUSIPCO 1998)*, pages 1–4. IEEE, 1998.

97. Nick G Kingsbury. The dual-tree complex wavelet transform: a new technique for shift invariance and directional filters. In *IEEE Digital Signal Processing Workshop*, volume 86, pages 120–131, 1998.

98. Norden E Huang, Zheng Shen, Steven R Long, Manli C Wu, Hsing H Shih, Quanan Zheng, Nai-Chyuan Yen, Chi Chao Tung, and Henry H Liu. The empirical mode decomposition and the Hilbert spectrum for nonlinear and non-stationary time series analysis. *Proceedings of the Royal Society of London. Series A: Mathematical, Physical and Engineering Sciences*, 454(1971):903–995, 1998.

99. Anand Parey and Ram Bilas Pachori. Variable cosine windowing of intrinsic mode functions: Application to gear fault diagnosis. *Measurement*, 45(3):415–426, 2012.

100. María E Torres, Marcelo A Colominas, Gaston Schlotthauer, and Patrick Flandrin. A complete ensemble empirical mode decomposition with adaptive noise. In *2011 IEEE International Conference on Acoustics, Speech and Signal Processing (ICASSP)*, pages 4144–4147. IEEE, 2011.

101. Zhaohua Wu and Norden E Huang. Ensemble empirical mode decomposition: a noise-assisted data analysis method. *Advances in Adaptive Data Analysis*, 1(01):1–41, 2009.

102. Konstantin Dragomiretskiy and Dominique Zosso. Variational mode decomposition. *IEEE Transactions on Signal Processing*, 62(3):531–544, 2013.

103. Jerome Gilles. Empirical wavelet transform. *IEEE Transactions on Signal Processing*, 61(16):3999–4010, 2013.

104. Abhijit Bhattacharyya, Lokesh Singh, and Ram Bilas Pachori. Fourier-Bessel series expansion based empirical wavelet transform for analysis of non-stationary signals. *Digital Signal Processing*, 78:185–196, 2018.

105. Pushpendra Singh, Shiv Dutt Joshi, Rakesh Kumar Patney, and Kaushik Saha. The Fourier decomposition method for nonlinear and non-stationary time series analysis. *Proceedings of the Royal Society A: Mathematical, Physical and Engineering Sciences*, 473(2199):20160871, 2017.

106. Binish Fatimah, Pushpendra Singh, Amit Singhal, and Ram Bilas Pachori. Detection of apnea events from ECG segments using Fourier decomposition method. *Biomedical Signal Processing and Control*, 61:102005, 2020.

107. Amit Singhal, Pushpendra Singh, Binish Fatimah, and Ram Bilas Pachori. An efficient removal of power-line interference and baseline wander from ECG signals by employing Fourier decomposition technique. *Biomedical Signal Processing and Control*, 57:101741, 2020.

108. Vipin Gupta and Ram Bilas Pachori. FBDM based time-frequency representation for sleep stages classification using EEG signals. *Biomedical Signal Processing and Control*, 64:102265, 2021.

109. Pooja Jain and Ram Bilas Pachori. Event-based method for instantaneous fundamental frequency estimation from voiced speech based on eigenvalue decomposition of the Hankel matrix. *IEEE/ACM Transactions on Audio, Speech, and Language Processing*, 22(10):1467–1482, 2014.

110. Pooja Jain and Ram Bilas Pachori. An iterative approach for decomposition of multi-component non-stationary signals based on eigenvalue decomposition of the Hankel matrix. *Journal of the Franklin Institute*, 352(10):4017–4044, 2015.

111. Rishi Raj Sharma and Ram Bilas Pachori. Time–frequency representation using IEVDHM–HT with application to classification of epileptic EEG signals. *IET Science, Measurement & Technology*, 12(1):72–82, 2018.

112. Steven L Brunton and J Nathan Kutz. *Data-Driven Science and Engineering: Machine Learning, Dynamical Systems, and Control*. Cambridge University Press, 2022.

113. Ozlem Karabiber Cura and Aydin Akan. Analysis of epileptic EEG signals by using dynamic mode decomposition and spectrum. *Biocybernetics and Biomedical Engineering*, 41(1):28–44, 2021.

114. J Nathan Kutz, Xing Fu, and Steven L Brunton. Multiresolution dynamic mode decomposition. *SIAM Journal on Applied Dynamical Systems*, 15(2):713–735, 2016.

115. Ervin Sejdić, Igor Djurović, and Jin Jiang. Time–frequency feature representation using energy concentration: An overview of recent advances. *Digital Signal Processing*, 19(1):153-183, 2009.

116. M Tanveer and Ram Bilas Pachori. *Machine Intelligence and Signal Analysis*. Vol. 748. Springer, 2018.

117. Dilip Singh Sisodia, Ram Bilas Pachori, and Lalit Garg. *Handbook of Research on Advancements of Artificial Intelligence in Healthcare Engineering*. IGI Global, 2020.

118. Varun Bajaj and Ram Bilas Pachori. Classification of seizure and nonseizure EEG signals using empirical mode decomposition. *IEEE Transactions on Information Technology in Biomedicine*, 16(6):1135–1142, 2011.

119. Mohit Kumar, Ram Bilas Pachori, and U Rajendra Acharya. Characterization of coronary artery disease using flexible analytic wavelet transform applied on ECG signals. *Biomedical Signal Processing and Control*, 31:301–308, 2017.

120. Pradeep Kumar Chaudhary and Ram Bilas Pachori. Automatic diagnosis of glaucoma using two-dimensional Fourier-Bessel series expansion based empirical wavelet transform. *Biomedical Signal Processing and Control*, 64:102237, 2021.

121. Pradeep Kumar Chaudhary and Ram Bilas Pachori. Automatic diagnosis of different grades of diabetic retinopathy and diabetic macular edema using 2-D-FBSE-FAWT. *IEEE Transactions on Instrumentation and Measurement*, 71:1–9, 2022.

122. Miao Zhang and Guo Wei. An integrated EMD adaptive threshold denoising method for reduction of noise in ECG. *PLoS One*, 15(7):e0235330, 2020.

123. Vladimir Ostojić, Tatjana Lončar-Turukalo, and Dragana Bajić. Empirical mode decomposition based real-time blood pressure delineation and quality assessment. In *Computing in Cardiology 2013*, pages 221–224. IEEE, 2013.

124. Iau-Quen Chung, Jen-Te Yu, and Wei-Chi Hu. Estimating heart rate and respiratory rate from a single lead electrocardiogram using ensemble empirical mode decomposition and spectral data fusion. *Sensors*, 21(4):1184, 2021.

125. Mohammod Abdul Motin, Chandan Kumar Karmakar, and Marimuthu Palaniswami. Ensemble empirical mode decomposition with principal component analysis: A novel approach for extracting respiratory rate and heart rate from photoplethysmographic signal. *IEEE Journal of Biomedical and Health Informatics*, 22(3):766–774, 2017.

126. Pramod Gaur, Ram Bilas Pachori, Hui Wang, and Girijesh Prasad. A multi-class EEG-based BCI classification using multivariate empirical mode decomposition based filtering and Riemannian geometry. *Expert Systems with Applications*, 95:201–211, 2018.

127. Pengpai Wang, Mingliang Wang, Yueying Zhou, Ziming Xu, and Daoqiang Zhang. Multiband decomposition and spectral discriminative analysis for motor imagery BCI via deep neural network. *Frontiers of Computer Science*, 16(5):1–13, 2022.

128. Xiongliang Xiao and Yuee Fang. Motor imagery EEG signal recognition using deep convolution neural network. *Frontiers in Neuroscience*, 15:312, 2021.

129. Dimitrios Dimitriadis, Petros Maragos, and Alexandros Potamianos. Robust AM-FM features for speech recognition. *IEEE Signal Processing Letters*, 12(9):621–624, 2005.

130. Hai Huang and Jiaqiang Pan. Speech pitch determination based on Hilbert-Huang transform. *Signal Processing*, 86(4):792–803, 2006.

131. Matteo Gandetto, Marco Guainazzo, and Carlo S Regazzoni. Use of time-frequency analysis and neural networks for mode identification in a wireless software-defined radio approach. *EURASIP Journal on Advances in Signal Processing*, 2004(12):1–13, 2004.

132. Pai Wang, Yongqing Wang, Ediz Cetin, Andrew G Dempster, and Siliang Wu. Time-frequency jammer mitigation based on Kalman filter for GNSS receivers. *IEEE Transactions on Aerospace and Electronic Systems*, 55(3):1561–1567, 2018.

133. Stuti Shukla, Sukumar Mishra, and Bhim Singh. Empirical-mode decomposition with Hilbert transform for power-quality assessment. *IEEE Transactions on Power Delivery*, 24(4):2159–2165, 2009.

134. Xianfeng Fan and Ming J Zuo. Gearbox fault detection using Hilbert and wavelet packet transform. *Mechanical Systems and Signal Processing*, 20(4):966–982, 2006.

135. Rui Ming Zhang, Xuefei Xu, and Donald G Truhlar. Observing intramolecular vibrational energy redistribution via the short-time Fourier transform. *The Journal of Physical Chemistry A*, 2022.

136. Erdal Dinç and Zehra Yazan. Wavelet transform-based UV spectroscopy for pharmaceutical analysis. *Frontiers in Chemistry*, 503, 2018.

137. Gonul Turhan-Sayan and Serdar Sayan. Use of time-frequency representations in the analysis of stock market data. In *Computational Methods in Decision-making, Economics and Finance*, pp. 429-453. Springer, 2002.

138. Jitka Poměnková and Roman Maršálek. Time and frequency domain in the business cycle structure. Agricultural Economics 58(7):332-346, 2012.

139. Sudipta Das. The time–frequency relationship between oil price, stock returns and exchange rate. *Journal of Business Cycle Research*, 17(2):129-149, 2021.

140. Dandi Jia, Qiang Gao, and Hui Deng. Stock market prediction based on time-frequency analysis and convolutional neural network. In *Journal of Physics: Conference Series*, volume 2224, page 012017. IOP Publishing, 2022.

141. Peterson Owusu Junior, Anokye M Adam, Emmanuel Asafo-Adjei, Ebenezer Boateng, Zulaiha Hamidu, and Eric Awotwe. Time-frequency domain analysis of investor fear and expectations in stock markets of BRIC economies. *Heliyon*, 7(10):e08211, 2021.

142. Song Cai, Lintao Liu, and Guocheng Wang. Short-term tidal level prediction using normal time-frequency transform. *Ocean Engineering*, 156:489–499, 2018.

143. Ravindra Pethiyagoda, Timothy J Moroney, Gregor J Macfarlane, Jonathan R Binns, and Scott W McCue. Time-frequency analysis of ship wave patterns in shallow water: modelling and experiments. *Ocean Engineering*, 158:123–131, 2018.

144. Moises Jimenez-Martinez. Fatigue of offshore structures: A review of statistical fatigue damage assessment for stochastic loadings. *International Journal of Fatigue*, 132:105327, 2020.

145. Kritiprasanna Das and Ram Bilas Pachori. Schizophrenia detection technique using multivariate iterative filtering and multichannel EEG signals. *Biomedical Signal Processing and Control*, 67:102525, 2021.

# Index

Printed in the United States
by Baker & Taylor Publisher Services